The Journey to Moonwalking

The People that Enabled Footprints on the Moon

The Journey to Moonwalking

The People that Enabled Footprints on the Moon

KENNETH S. THOMAS

CP
Curtis Press

Kenneth S. Thomas

ISBN: 978-0-9934002-2-3

© Kenneth S. Thomas 2017

This trade edition published in 2019 by Curtis Press

Cover design: Chandler Book Design, Kings Lynn

Distributed in North America by SCB

Visit Curtis Press at *www.curtis-press.com*

Contents

Acknowledgments

While it is impossible to acknowledge all who have provided support and contributions to this book over the decades, the author wishes to thank:

Earl Bahl (former Hamilton Standard a.k.a. "Hamilton" or "HS" Apollo Engineer and Mechanical Design Manager), Jack Bassick (retired David Clark Executive Vice President and Historian), Ronald "Ron" Bessette (former International Latex Corporation/ILC-Dover a.k.a. "ILC" Liquid Cooling Garment Engineer), William "Bill" Bouchelle (former Hamilton Apollo Engineer), Edgar H. "Ed" Brisson (former Hamilton Apollo Engineer), Harlan Brose, (former Hamilton Apollo Engineer), Ray Dudek (former Hamilton Apollo Helmet Worker), S. Victor "Vic" Fleischer (Associate Professor University of Akron & B.F. Goodrich Historian), Dennis Gilliam (Spacesuit Design Engineer and Historian), Jerry Goodman (former National Aeronautics and Space Administration a.k.a. "NASA" Apollo Pressure Suit Manager, 1962-1965), Kenneth "Ken" Griffin (former Hamilton Apollo Helmet Engineer), Gary L. Harris (Spacesuit Engineer, Author, and Historian), Andy Hoffman (former Hamilton Apollo Spacesuit Project Engineer), ILC-Dover Corporation, David Jennings (retired Hamilton Engineer and inventor of the Apollo Liquid Cooling Garment), John "Jack" Kelly (retired Hamilton Space Engineer/Manager), Kennedy Space Center Aerospace Education Department, Joe Kosmo (former NASA Advanced Spacesuit Manager 1962–1968), Lilian Kozloski (retired National Air and Space Museum Spacesuit Specialist and Author), Douglas Landry (Historian and Apollo Spacesuit Researcher), Dick Lawyer (former Manned Orbiting Laboratory Pilot and Program Suit Subject/Evaluator), Dr. Cathleen Lewis (National Air and Space Museum Space History Specialist), William "Bill" Maas (ILC Field Office), James "Jim" McBarron II (former NASA Apollo Pressure Suit Manager 1966-1972), Daniel "Danny" McFarlin (ILC Field Office), Harald Joseph "Joe" McMann (former NASA Apollo Suit Subject 1962–1972, Skylab Suit Manager 1969–1974, and co-author of US Spacesuits), Michael A. "Mike" Marroni Jr. (former Hamilton Apollo Engineer), James "Jim" O'Kane (NASA retired), William "Uncle Bill" Rademakers (former Hamilton Apollo Engineer), Donald "Dr. Flush" Rethke (former Apollo engineer), Dixie Rinehart (former ILC Apollo Glove Engineer), Thomas Sanzone (former Hamilton Apollo Engineer), Bob Scagni (Hamilton retired), Tom Schuetz (Hamilton retired), David H. "Dave" Slack (former ILC Apollo Reliability Engineer), Betty Slack (Dave's widow), Roger Weatherbee (former Hamilton Apollo Manager), Dick Wilde (former Hamilton Apollo Engineer), and Amanda "Mandy" Young (retired National Air and Space Museum Apollo Spacesuit Specialist).

Note: For abbreviations see Appendix II—Acronyms.

Acknowledgements

Foreword

Most people's earliest memories of spacesuits are the grainy, black and white images of a man on the Moon, wearing a "puffy" white suit with an oversized, gold-visored helmet and a large backpack, who was walking, hopping, or driving around on the lunar surface. How these suits and the men who wore them reached the Moon is a superb story of imagination, fortitude, individualism, and courage. However, the design and construction of these spacesuits is a story of equal fascination, imagination, and determination. The people who designed and built these wonderful suits had no previous models from which to work. The materials and design had to be very different from the Mercury and Gemini models, and had to perform different functions. The materials had, in the majority of cases, to be invented or redesigned. The suits themselves had to be designed, constructed, fitted, and tested in an amazingly short period of time—it was only 8 years from John F. Kennedy's mandate to Neil Armstrong's walk on the Moon, which was in turn only about 30 years from the first simple but usable, fully pressurized suit.

The "space race" was a follow-along from the "high-altitude race" that had been progressing since the end of World War II. By the end of the 1950s the "race" had developed into major tensions between the United States and Soviet Union, with Cuba in the middle, and was verging on armed conflict. The race between the U.S. and the U.S.S.R. to be the first nation on the Moon redirected some of this energy into a "space technology competition," with the U.S. landing men on the Moon in July 1969, finally reaching U.S./U.S.S.R. cooperation with the Apollo-Soyuz Test Operation and the "handshake in space."

This book is the story of some of the people involved, their ingenious ideas, willingness and necessity to "think outside the box," their imagination, and hard work. It is about the companies for whom these people worked and who were willing to "run" with their sometimes extremely unconventional ideas, and it is about NASA, who needed to bring it all together to succeed with this enormous task. All the while never forgetting that behind all the decisions and schedules was the pressure of the mandate to succeed before the Russians did.

To achieve John F. Kennedy's 1961 mandate of "landing a man on the Moon and returning him safely to the Earth," it was obvious that it would not be possible without both redesigning spacecraft and spacesuits. At this time, spacesuits were designed to provide life support for the astronaut only while he was sitting in, or tethered to, the spacecraft, and there were none capable of allowing an astronaut to explore the lunar surface freely. While walking on the Moon, it would be impossible for the astronaut to remain attached to the spacecraft, requiring the design and manufacture of a spacesuit capable of permitting independent movement. To design and build such a spacesuit represented a formidable task, particularly in the timeframe mandated.

It would have been absolutely impossible for the astronauts to have walked on the Moon without a spacesuit. The Apollo spacesuit is the smallest and most flexible of spacecraft, and the man inside had to be protected from the most hostile of environments, be able to walk, bend, talk, breathe, stay warm or cool, drink, and go to the bathroom, all while sealed in this suit. To construct such a vehicle required the combined efforts of men and women in all parts of the United Statesscientists, engineers, seamstresses, designers, and technicians, to name but a few. The project needed people to design and build the suit, gloves, boots, and helmets themselves, along with the breathing apparatus, including oxygen tanks, hoses, and a mechanism for "scrubbing" the air. It needed people to design and build communications and waste disposal systems while adhering to the requirement of low weight, bulk, and flexibility. Above all, the astronaut had to be able to move with some degree of comfort. These requirements and tasks were of truly Herculean proportions, but it was done— these people did it, and there was never a suit failure. By the time Neil Armstrong walked on the Moon, the spacesuit he wore was made of 26 layers of modern materials and textiles, with glues, metals, and plastics, along with a highly sophisticated breathing and heating/cooling system that weighed roughly 125 pounds.

Though much of the documentation pertaining to the Apollo program has been lost, and sadly many of the people involved have either died or moved on, there is much to be gleaned from memories, notes, letters, and other documents. The spacesuit community has always been a small one, and the historians are now usually willing to share any information they may have. Nevertheless, it still required considerable research and patience on the part of the author to put this book together, however, the rewards are great—for both history and space-enthusiast communities.

The story of the men and women who embarked on this quest is one that has never been told before. The subject of spacesuit history, design, and construction is so vast that it is a life's work for historians and scholars, and is almost impossible to compile into one book. Consequently, books on the subject have focused previously on the completed spacesuits, their uses, histories, and the missions for which they were made, along with the differences between various models. Nonetheless, these stories needed to be told, and are very important and captivating. However, there is also a fascinating historical interest in the individuals who worked behind the scenes building this incredible equipment—that actually made the voyages to the Moon possible—and their history also needs to be explored. This book tells a story that is just as important and just as interesting as many of the "other" stories.

Amanda Young
National Air and Space Museum Museum Specialist,
Astronaut Equipment, Retired
Author of
Spacesuits, The Smithsonian National Air and
Space Museum Collection—2009

Introduction

This is a story of people who struggled and sacrificed to make possible one of the greatest human achievements of the 20th century. As these people were part of the United States' space effort, this could be viewed as an American story. However, the space race was a rivalry between the superpowers where otherwise ordinary men and women provided a positive influence on the course of world history at a time when it was desperately needed. Thus, it is also a human tale common to all humanity.

The Apollo spacesuit made history, and that history changed the world. However, most people do not understand the importance of spacesuits. If your spacecraft loses cabin pressure while going into space, while in celestial travel, or while returning to Earth, your spacesuit can keep you alive. If there is a problem with your spacecraft or space station, your spacesuit can allow you to go out into space and address a potentially life-threatening emergency. Even going out into space to conduct planned activities can have great significance. However, perhaps the greatest potential value is that a spacesuit can make it possible to venture onto planetary bodies and explore the secrets of the universe.

The goal of the space race was not only to go to the Moon but to conduct meaningful exploration once there. This required leaving the spacecraft and performing tasks on the lunar surface. At the start of this competition, no one had spacesuits capable of allowing people to venture outside a spacecraft, let alone explore the surface of the Moon. The craft and technologies to make such spacesuits were yet unknown. While some of the challenges that lay ahead were recognized, most were not. To appreciate the significance of contributions made by people working on spacesuits it is necessary to understand the challenges they faced.

For people to survive and function, they must be surrounded by a pressurized environment that contains sufficient oxygen. To work on the surface of the Moon, this meant being in a pressure suit. The first challenge comes from the pressure in the suit trying to make the pressure garment immobile. As a minimum, this suit environment had to be pressurized to at least 3.5 pounds per square inch (3.5 psi or 24 kPa) pure oxygen for an astronaut to effectively function in the vacuum of space. A typical pressure suit has over 1,000 square inches (6,452 cm^2) of internal surface area. Thus, there are thousands of pounds of pressure trying to make the garment rigid and inflexible. Without effective mobility elements, the astronauts would be unable to move their arms, legs, waist, and fingers adequately enough to explore the Moon's surface. Additionally, the pressure suit had to bend so it would follow their movement. Otherwise, the garment would restrict movement and cause injury. This required development and invention.

In the direct sunlight of space, the temperature of exposed items on the lunar surface can rise to 250°F (121°C). In the shade of space, things cool to approximately −140°F (−96°C). Fortunately for spacefarers, highly effective space insulation was developed in the early 1960s. However, the success of this insulation caused another problem; it held both body heat and heat generated by equipment in the spacesuit. The life support system not only had to provide pressure and oxygen but also a cool working environment.

Rejecting heat from an insulated spacesuit is not easy. First, the spacesuit must remove heat from the user so that they remain comfortable and can effectively work. Pressure suits have to be reasonably tight fitting. This is not conducive to heat removal using circulating gases. Then, the suit's life support, a.k.a. "backpack," must collect the heat generated from the user and other sources, and reject all the collected heat to the vacuum of space. Of course, a vacuum is highly effective insulationthere are no particles to carry away the heat energy.

The life support backpack also had to remove carbon dioxide so headaches, disorientation, and loss of life did not occur. The backpack provided oxygen to sustain life, and controlled humidity for both comfort and safety. All these backpack functions were provided in a system so compact that it and the pressure suit could pass through hatches and be light enough to be carried in lunar gravity for hours. Moreover, there were basic life considerations such as staying hydrated and "going to the bathroom" that had to be addressed.

However, the greatest spacesuit challenges were unknown. No one knew the correct design requirements as no one had ever designed, successfully manufactured, or certified the safety of a spacesuit capable of going out in space before. Then there was developing and making the spacesuits that would meet those requirements and perform well on the Moon. All these issues were great challenges to Apollo.

The journey that enabled humankind's first footsteps on the moon is a collage of human experiences spanning over three decades. The "giant leap for mankind" was made possible by hundreds of small steps. Each step was a challenge. When the challenge was underestimated, a setback resulted. Thus, this story is of scores of iterations of effort by hundreds of people from a variety of organizations. Through perseverance and dedication, an army of workers all made contributions culminating in the spacesuit that made this historic feat possible. These contributions not only helped shaped world history but continue to leave influences as humanity advances toward its destiny of moving beyond our home planet in the future.

CHAPTER 1

A Garden Epiphany and the Path to the Apollo Competition

It started in an Ohio garden on a hot summer's day in 1943. An exhausted engineer named Russell Colley was seeking distraction from the burden of a nation struggling to turn the tide of a world at war. He was a passionate explorer of early pressure suit technology. His previous iterations of wartime suit prototypes were not mobile enough when pressurized. He knew the expectations placed upon him were high. The United States had suffered great losses in the Pacific and North Africa in the preceding year. The ability for bomber crews to fly high above enemy aircraft and ground defenses could be key to changing the direction of the war, but why would people think this man might have the solutions to these challenges?

Nine years earlier, Colley had developed the world's first successful high-altitude pressure suit for a daredevil pilot named Wiley Post (Figure 1.1).

Post recognized the west-to-east speed advantage using high-altitude winds which are now called the jet stream. He also recognized the perils. Without a pressurized enclosure, he would perish long before he reached the required altitude. Once there, he would need an effective heating system to protect him from lower-than-Arctic temperatures. Post knew that a pressurized cabin was not an option as a result of its weight and development costs. He had to devise a system that would work with his existing aircraft, a single-engine Lockheed Vega aircraft named *Winnie Mae*. Post knew how to provide the needed life support but lacked a fabric suit to retain pressure while allowing him to fly his airplane. For that, Post hired B.F. Goodrich.

The path that caused Colley to be the supporting Goodrich engineer was probably unique. Born in Stoneham, Massachusetts in 1899, Colley originally wanted to design women's clothing. However, this was seen as an "unmanly" profession and he was urged to enroll in a technical college instead. After graduation in 1928, Colley moved to Akron, Ohio to become an engineer for Goodrich. There, he supported many products including those for aircraft. In the process, he created concepts or took the ideas of others to make functioning products.

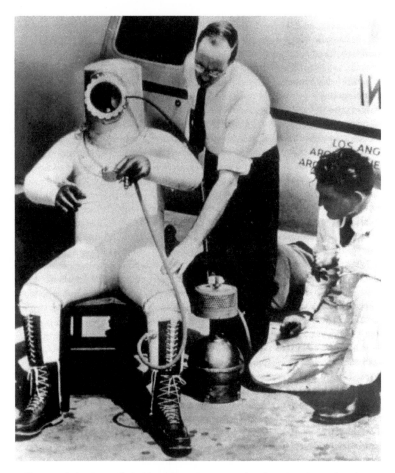

Figure 1.1. Russell Colley (center) supporting Post's suit evaluation
(courtesy Gary L. Harris)

In March 1934 Colley was assigned to be the engineer on Wiley Post's suit system project. The challenges were immense. The suit had to withstand a 3-psi internal pressure without failure. However, at that pressure the garment became so stiff that the person inside could barely move, let alone fly an airplane. It took Colley three prototype suit efforts before being successful (Figure 1.1). On September 5, 1934, suit "Number 3" supported flight at 42,000 ft. Subsequent flights to and over 50,000 ft were made, catching public attention which included Post/Goodrich suits being used as a key part of the 1935 science fiction movie, *Air Hawks*, the heroes successfully flew over a villain's death ray. While the cinematography and technologies were crude by today's standards, the movie was the ultimate in the science fiction genre of the time. These exploits made Colley a celebrity within the aviation industry. Thus, expectations on him were high.

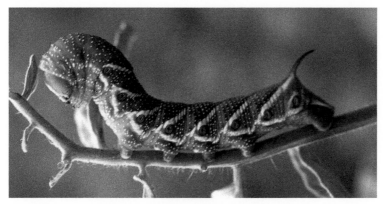

Figure 1.2. The Tomato Hornworm
(courtesy Ward Upham)

Back home in the garden, a camouflaged, almost invisible tomato worm (a tomato hornworm larva species, *Manduca quinquemaculata*) caught Colley's eye. He stopped to watch it as it moved on a twig (Figure 1.2). Its convoluted body flowed effortlessly as it moved. He wondered, could the convolutes be the secret to how it moved and changed shape with ease?

Colley soon devised a molded convolute mobility joint of dipped fabric, cords, and cables that was incorporated into a pressure suit prototype (Figure 1.3).

Alas, the B.F. Goodrich XH-5 "Tomato Worm Suit" was a significant improvement but still not the pressure suit solution the military was seeking. The joints provided mobility only in a back-and-forth direction as the joints pivoted on their side cables like a swing in a playground. World War II (WWII) came to a close without the desired mobility breakthrough. However, this basic "molded rubber" mobility joint design, with subsequent improvements, was destined to be a feature of Apollo pressure suits, thus beginning the technology trail that concluded with men exploring the Moon.

The molded rubber mobility joint was not the only Apollo pressure suit technology that was rooted in this U.S. WWII high-altitude suit development program. The Army Air Corp elected to share developments with all contractors creating a cross-pollination of ideas and concepts. As a result, the program also contributed the use of mobility elements that slid on cables, which enjoyed popularity with Goodyear (Figure 1.4) designs among other contractors. While these early attempts introduced features that would later appear in Apollo spacesuits, they required excessive effort to achieve mobility. The art of reducing friction was yet to be achieved. Pressure suit development would continue after WWII.

These wizards of structural fabrics worked their spells during WWII on inflatables such as rafts and in lighter-than-air craft such as blimps (Figure 1.5). What made blimps so amazing is that with the exception of the gondola section

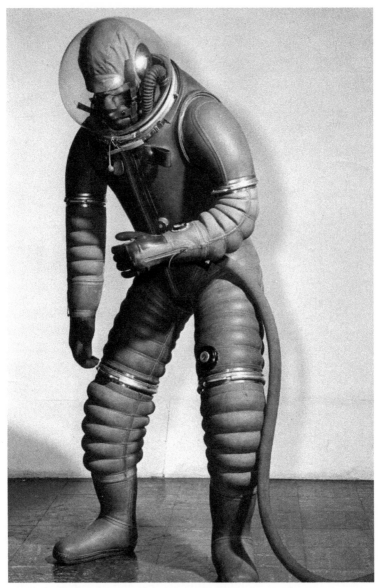

Figure 1.3. The Goodrich XH-5 Tomato Worm Suit
(courtesy Gary L. Harris)

that hung on the underside, virtually the rest of this enormous aircraft was fabric that was under pressure. Yet, these blimps were durable enough to reliably resist most types of violent weather. However, developments during the war were for national survival. Any organization that could contribute to the effort did so willingly. After the war, industry transitioned to normal business

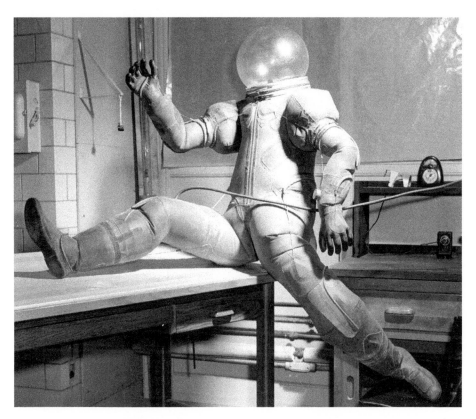

Figure 1.4. The Goodyear XH-9 prototype
(copyright courtesy Gary L. Harris)

Figure 1.5. A World War II Navy Blimp
(courtesy U.S. Navy)

activities focused on profits. Most of the organizations that supported WWII pressure suit developments did not continue in this technical specialty.

Going into Apollo, there were many who dreamed that they or their organizations would be the creators of the spacesuit used on the Moon. This resulted in many organizations competing for this honor. For Mercury and Mercury Mark II, which was renamed Project Gemini in 1962, NASA elected to be both the planner and integrator of the spacesuits. Apollo would be different. However, NASA was a relatively new organization that was struggling to assemble the resources needed to support the growing space race. With two manned space programs running in parallel, NASA management believed they did not have the ability to handle this level of responsibility over a third and more demanding spacesuit effort that was needed for Apollo. This decision not only meant the need to fill a great variety of supplier roles but also hold a competition for a prime contractor to head this spacesuit effort on behalf of NASA. Fortunately, the U.S. space program had an abundance of qualified competitors. Perhaps the most experienced of these competitors was B.F. Goodrich.

In 1869, Benjamin Franklin Goodrich purchased the Hudson River Rubber Company, which he relocated to Akron, Ohio. Renamed B.F. Goodrich Company, this enterprise started producing fire-fighting hoses, but soon branched out to selling garden hoses and bicycle tires. In the 1890s, the company introduced a pneumatic tire, initially for bicycles and later for automobiles.

B.F. Goodrich was not afraid to try new products and supported Wiley Post's attempts to set aircraft speed records, at high altitudes, in the mid-1930s. In WWII, Goodrich was one of many organizations that supported the nation's top secret MX-117 pressure suit program, of which the Tomato Worm Suit was but one of the Goodrich prototypes.

In 1947, the Air Force and the Navy agreed to pursue different approaches to high-altitude suits. The Air Force specialized in developing partial-pressure suits which were typically a two-part system where pressure was applied to the body and the airman breathed oxygen through a mask (Figure 1.8, see p. 10). The Navy focused on full-pressure suits. Goodrich produced a series of prototype designs for this effort.

One suit in this series was the Omni-Environment Suit (U.S. Patent 2,966,155, Carroll P. Krupp inventor). Outside of work, Krupp was a family man, builder, and craftsman. Married for 63 years, he and his wife Irene had six children, all daughters. Krupp was always making something such as furniture for a family member or a competition-winning model airplane. One time, he designed and built a car with a custom fiberglass body he fabricated himself. At Goodrich, he rose from the ranks with just a high school diploma to be a non-degreed engineer through hard work and a talent for practical ideas. During his career, he received over 40 patents for a variety of products ranging from commercial products to high-altitude pressure suits, the precursors of spacesuits.

The Omni-Environment Suit featured more advanced molded rubber

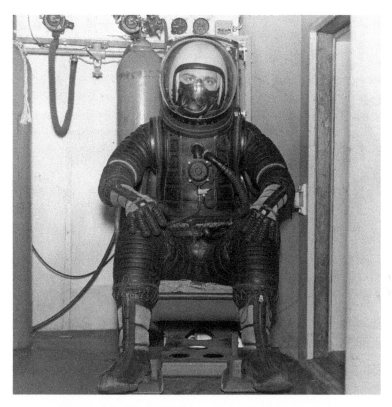

Figure 1.6. The 1953 Krupp Omni-Environment Suit
(courtesy the University of Akron Archival Services)

convolute joints at the shoulders, elbows, and knees. Like Colley's Tomato Worm Suit, the joints of Krupp's suit provided mobility back and forth. While its mobility-joint construction broke from the Tomato Worm and lunar suit path, the Omni-Environment Suit contained a key feature of the suits used to explore the surface of the Moon as it had upper-arm, pressure-sealed bearings located at the biceps (Figure 1.6). This permitted the elbow's one axis of movement to rotate allowing movement in any direction. This would be key to lunar and subsequent space exploration.

This Omni-Environment Suit was featured on the February 28, 1953 cover of *Collier's* magazine under the title "World's First Space Suit." The article included noted scientists such as Drs. James Van Allen and Wernher von Braun who argued the case that the U.S. essentially had the existing technology needed to put men in space.

In 1956 Goodrich developed its Navy Mark II Suit. In the context of the Navy full-pressure suit development program, this was but a passing design iteration. However, this was a further step in the technical evolution that influenced lunar suit development. Pressure-sealed bearings were provided at the shoulder, mid-upper arm, and forearm, which resulted in a revolutionary improvement in arm mobility. To demonstrate this accomplishment to the public, the Navy allowed *Life* magazine to feature suit users exercising on a rowing machine, riding a bicycle, playing baseball, performing gymnastics on parallel bars, dissecting biology specimens, and assembling a TV antenna while pressurized.

However, in the context of military aviation, especially for fighter pilots, this was not a success. The cockpits of fighter aircraft are designed to tightly fit around the pilot to minimize volume to allow faster air speeds. In an evasive maneuver, the gravitational forces on a 180-pound pilot can easily be over a ton (907 kg). Under those loads, the pilot could be injured from being pushed against the hard metal bearings. So this approach to pressurized mobility came and went in the U.S. military's quest for airmen survival. However, the Mark II arm and bearing concept was not lost. It would return later in Apollo's spacesuit development.

With the Goodrich Mark III design, the shoulder and upper-arm bearings were removed taking this aviation evolution away from the Apollo path. The resulting performance in aviation use caused Goodrich to become the Navy's supplier of choice. A lighter weight, more comfortable Goodrich Mark IV soon followed. Ironically, this positioned the company to become the spacesuit supplier to America's first space program, Project Mercury because NASA wished to take an existing, well proven design and adapt it for its first manned program but proved to be an impediment to consideration for Apollo.

On January 29, 1959, NASA convened its first spacesuit conference with the nation's leading experts in attendance. This included three competitors: the B.F. Goodrich Company which supplied most of the pressure suits used by the Navy; the David Clark Company which was under contract to the Air Force for X-15 pressure suits; and the International Latex Corporation of Dover, Delaware, which was a helmet provider to the Air Force for flight suits and had been an X-15 pressure suit competitor.

NASA's goal was to select the suit design and designing/supplying organization for its Mercury program. Goodrich was selected and finally awarded the contract for the Mercury spacesuit on July 22, 1959. Russell Colley, along with Carl Effler, Donald Ewing, and other Goodrich employees were key to the adaptation of the Navy Mark IV pressure suit to meet NASA's needs for orbital space flight safety. For Mercury, Goodrich essentially added an aluminized coating to the outer layer and extra entry zippers (Figure 1.7).

While Mercury suits did not allow an astronaut to perform activities outside the spacecraft, they shared a common problem with Apollo in that astronauts could be required to wear their suits in a pressurized condition for many hours during which the astronauts would need to drink. The solution was the "Helmet with Pressure Seal (Drink) Port" (U.S. Patent 3,067,425, Russell S. Colley inventor). This design was used not only for Mercury but also for Gemini and Apollo. Thus, Goodrich personnel left their fingerprints on the suits used on the Moon.

Figure 1.7. 1963 Goodrich-provided Mercury suits
(courtesy NASA)

There was another U.S. industry that would play an even greater founding role in Apollo pressure suits. This industry was women's undergarments, specifically starting with girdles. While girdles are garments that have generally passed into history, such items were at one time essential parts of women's wardrobes. For centuries, corsets were part of Western civilization's female formal attire, creating slimmer, "hour glass" waists. In the U.S., corsets essentially ended with the U.S. entry into World War I in 1917. The U.S. War Industries Board asked women to stop buying corsets to free up metal for production required for the war. The replacement was the girdle, an all-fabric structure, which remained a prominent part of female wardrobes for another half century.

In 1935 David M. Clark founded a company to manufacture women's "snug fitting garments" (more commonly called girdles) along with other undergarments. With the outbreak of WWII, the David Clark Company became part of the war effort developing and manufacturing anti-gravitation (anti-G) suits, consisting of bladders, coveralls, and controls to allow fighter pilots to remain functional under extreme centrifugal forces from aerial maneuvers that would otherwise render the pilot unconscious.

Figure 1.8. James Henry and the S-1 suit
(courtesy G. Harris)

After WWII, this path continued with the development of a partial-pressure (mechanical counter-pressure) suit system for high-altitude aircraft. This started with the collaboration of Dr. James P. Henry of the University of Southern California and David M. Clark, founder of the David Clark Company.

Henry conceived a system in which a gas mask provided pressurized oxygen to the face and lungs. The gas pressure also expanded "capstan-tubes" running down the torso, arms, and legs, which tightened the garment providing external counter-pressure to the torso (Figure 1.8). Clark provided resources and technical support from his Worcester, Massachusetts facilities to Henry. The resulting prototype was tested to an equivalent of 90,000 feet (27 km) at Wright Field, Ohio in 1946. Clark then developed Henry's design into the S-1 flight suit used during flights of the supersonic Bell X-1.

The X-1 plane was followed by another National Advisory Committee on Aeronautics (NACA) rocket plane program named D-558-2 Douglas Skyrocket. This had the goal of going faster than twice the speed of sound (i.e., breaking Mach II). For this, a better pressure suit system was sought. In 1951, the David Clark Company won the D-558-2 suit development contract, which was their first operational, full-pressure suit (fully enclosed) design. The success of this suit system established David Clark as a credible full-pressure suit competitor.

In 1954 NACA joined the U.S. Air Force and Navy in a joint experimental aircraft/spacecraft named the X-15. The X-15 program's mission was to expand significantly the horizons of aerospace research by operating as an aircraft at many times the speed of sound. However, the X-15 was designed to resist the heat and friction of atmospheric reentry and was powered by rocket engines that were not dependent on air for propulsion because the X-15 was also intended to be a suborbital space-plane. The design called for full-pressure suits in case the cabin of the X-15 lost pressure. As the rocket-plane flew into and returned from space, X-15 pressure suits were also used as in-vehicle spacesuits.

The selection and development of X-15 suits started when the Air Force invited several companies to provide pressure suit designs for consideration. Prototypes from the David Clark Company, the Rand Corporation, and the International Latex Corporation were among the suits funded by and evaluated at Wright-Patterson Air Force Base in Ohio in 1957. The winner of the X-15 suit evaluation and subsequent contract was the David Clark Company Suit (Figure 1.9). This positioned the David Clark Company as a serious pressure suit competitor for Apollo.

The International Latex Corporation was founded in 1932 by Abram Nathaniel (A.N.) Spanel to produce many of his inventions using latex rubber.

Figure 1.9. David Clark X-15 suits circa 1963
(courtesy NASA)

Figure 1.10. The International Latex K-2 Air Force Helmet
(courtesy ILC-Dover LP)*

These inventions included form-fitting baby diapers and shower caps. Operations were moved to Dover, Delaware starting in 1937. In Dover the product line expanded to include women's Playtex® undergarments, which quickly grew to be the corporation's best-selling product line and an industry leader. However, International Latex continued to pursue other forms of business to become a diverse organization.

WWII brought International Latex's first foray into pressurized structural fabrics with the manufacture of inflatable rafts. Following the war, International Latex's Specialty Unit looked to expand into high-altitude oxygen masks for the Defense Department. In 1952 International Latex was awarded a contract to supply the Navy and Air Force with high-altitude pressure helmets (Figure 1.10).

In 1955 International Latex's Specialty Unit was renamed the Special Products Division, which soon looked to expand into high-altitude pressure suits. To support this product expansion, the Special Products Division recruited a talented, multifaceted individual named George P. Durney. Durney learned to fly as a teenager in Dover, Delaware and joined the Army Air Corps at the outbreak of World War II. He emerged from the war a decorated combat veteran. After the war, Durney earned a living flying crop-dusting aircraft in California. There he became concerned about human exposure to chemicals. Durney entered University of California—Los Angeles and took courses in materials and design with the goal of creating protective flight suits. However, before reaching this goal, he returned to Delaware to join International Latex.

The Air Force invited several companies to provide designs for the X-15 pressure suit for consideration. Prototypes from the David Clark Company, the Rand Corporation, and International Latex were among the suits funded by, and evaluated at, Wright-Patterson Air Force Base in Ohio in 1957. International Latex was funded to provide a pressure suit prototype for evaluation. For this, International Latex combined internal resources with outside recruitment which included personnel from B.F. Goodrich. This appears to have resulted in a cross-pollination of experience. While George Durney had no previous experience with B.F. Goodrich technology before joining International Latex, he understood why the molded convolute joint approach that was pioneered by Goodrich in World War II (Figure 1.3) did not reach its potential. Using the same materials and base construction methods but with revised geometries, Durney literally invented (U.S. Patent 3,432,860) a more effective mobility system. This was first embodied in International Latex's X-15 prototype (Figure 1.11). This mobility design would be a key feature of all International Latex and later ILC-Dover pressure suits for the Apollo, Skylab, and Apollo-Soyuz Test Project programs. Durney would continue to accrue Apollo patent credits to become the single greatest contributor to the Apollo spacesuit.

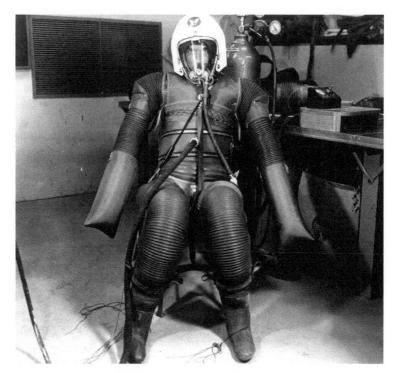

Figure 1.11. The International Latex X-15 prototype
(courtesy ILC-Dover LP)*

A curious feature of the X-15 prototype was that it was not equipped with pressure gloves (Figure 1.11) as the evaluation was for mobility systems and providing pressure gloves with the suit prototype was not a requirement. While the International Latex prototype was not a finalist in the X-15 suit competition, it did demonstrate to the fledgling space community that International Latex had become a competitor.

Project Mercury did not have a formal competition by later contracting standards. Instead, NASA selected a few recognized potential suppliers and funded them on a sole source basis to provide prototypes for evaluation. International Latex was one of the contractors selected to provide its latest technology prototype for evaluation. This suit had gloves. While International Latex was again not selected as the program's suit provider, it incrementally advanced International Latex's credibility as a potential Apollo suit provider.

International Latex's preparation for what would be the Apollo Space Suit Assembly competition started without hesitation following the end of the Mercury competition in 1959. International Latex knew NASA would need extravehicular activity (EVA) capacity and looked for opportunities to demonstrate its vision to meet that need (Figure 1.12).

International Latex recognized that NASA would be looking for a complete spacesuit system provider. This required being able to offer a life support capacity. In the period 1959-1960, International Latex started working with Garrett's AiReseach Division on a spacesuit joint venture focused on lunar exploration. In 1961 International Latex ended the collaboration to team up with Westinghouse and Republic Aviation, thus putting in place the organizational structure to support International Latex's 1962 Apollo competition proposal.

Yet another Apollo competitor, Hamilton Standard, was best known for airplane propellers. In 1909, Thomas F. Hamilton co-founded a company in Seattle, Washington that made wooden airplane propellers. The company soon became a success and a significant manufacturer of (wooden) propellers for the aviation industry. Separately, the Standard Steel Propeller Company was founded in Pittsburgh, Pennsylvania in 1919. This company developed the ground adjustable (metal) propeller that was used by Charles Lindbergh to fly non-stop from New York to Paris in 1927. By 1929 the United Aircraft & Transport Corporation acquired both the Hamilton and Standard Steel propeller companies. In 1932 the two companies were merged to form the Hamilton Standard Propeller Division, which was relocated to East Hartford, Connecticut. Hamilton subsequently developed the in-flight variable pitch propeller that was used on virtually all fighter and bomber aircraft of all nations in World War II. In recognition of contributions as a government contractor, the *Memphis Belle* and her crew flew into the Hamilton facility in East Hartford for a visit (Figure 1.13). The *Memphis Belle* flew 29 combat missions and was recognized as the first U.S. bomber to fly 25 missions without loss or injury to a crewmember.

In 1949, Hamilton broadened its product line to include jet fuel controls.

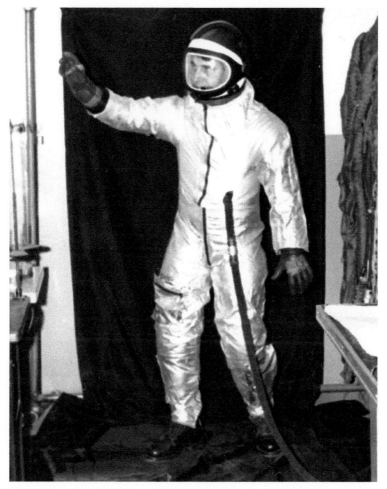

Figure 1.12. George Durney and the SPD-117 suit in a circa 1960 promotional
(courtesy ILC-Dover LP)*

Soon after, Hamilton drew on the same technology base to include aircraft environmental control systems. The aviation product lines introduced fiberglass manufacturing that brought with it structural fabric manufacturing disciplines that later played a role in the Apollo program.

At the start of 1959, Hamilton initiated funding of research and development in many space life support technologies. This included setting aside a staffed laboratory for space life support development. Since the development was principally focused on lunar exploration, the lab quickly earned the name the "Moon Room." Some of the key research and development efforts were the Centrifugal Water Separator and the lithium hydroxide (LiOH) cartridge for carbon dioxide removal that were later used on the Moon along with a

Figure 1.13. *Memphis Belle* and crew in East Hartford in 1944
(courtesy UTC Aerospace Systems)

pioneering Molecular Sieve Regenerable Carbon Dioxide Removal System (Figure 1.14).

Hamilton's first exposure to pressure suits came as they conducted studies during the preparation of a proposal in 1960 for the Oxygen Supply System for the then Mercury Mark II Astronaut Maneuvering Unit. That year, Hamilton won Mercury Mark II contracts. Among these was the development of the Oxygen Supply System. In 1962 Mercury Mark II was renamed and continued as Project Gemini. This gave Hamilton exposure to the latest B.F. Goodrich and Arrowhead Rubber pressure suit technologies but no first-hand experience. In 1960 NASA had only one active manned space program, Mercury, for which B.F. Goodrich was the pressure suit provider. As a result of hardware limitations and schedule constraints, gaining access to Mercury suits (Figure 1.7) was extremely difficult. However, the restraint and mobility joint systems of the Mercury spacesuits were the same as the then current Goodrich Navy high-altitude pressure suits. Through the U.S. Navy, Hamilton's Elliot Rosenthal arranged the on-site study of naval suits where Hamilton engineers were able to examine and test them (Figure 1.15).

During this early period, Goodrich and Arrowhead were considered the top contenders. However, Hamilton's study of the pressure suit industry indicated the David Clark Company was also a qualified suit provider and potentially receptive to teaming. In 1961 Hamilton entered into negotiations for David Clark pressure suit support of Hamilton's subsequent Apollo Space Suit Assem-

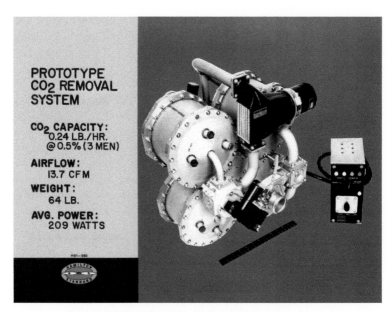

Figure 1.14. The 1960 Molecular Sieve prototype
(courtesy UTC Aerospace Systems)

Figure 1.15. Hamilton Standard evaluating a Navy pressure suit
(courtesy UTC Aerospace Systems)

bly proposal. By the year's end, a support agreement had been reached. In early 1962 David Clark brought its most recent prototype (Figure 1.19, see p. 25) to Hamilton for demonstration and evaluation. Thus, the stage was set for Hamilton to submit, as a potential spacesuit system provider, its 1962 Apollo competition proposal.

The aforementioned competitors may have had different product experiences but in these diverse products there was a common link, a shared artistry of structural fabrics that included dimensional control, thread spacing in seams, and selection of fabrics. This dimensional control, be it for blimp pressure bodies or bras, established a discipline that would be taken to a new level in later pressure suits to first gain and then repeat the desired performance. Controlling the spacing of the thread in seams that joined suit sections would be critical. If the spacing was far apart, too much load would be placed on the thread resulting in a weak joint and the thread potentially failing when the garment becomes loaded. If the spacing was too close together, damage may be caused to the fibers in the fabrics resulting in the fabrics being weakened, and ultimately failing.

The optimum relationship for maximum capacity in pressure suits was established given the characteristics of the fabrics and threads selected, and then the spacing was tightly controlled. With the selection and control of the materials, the artistry of spacesuits ventured into an almost magical realm. Before space exploration, there was no way to calculate the loads on complex shapes such as pressure suits. Additionally, fabrics were not made for space applications. They were made for commercial applications and selected for potential space use. There was no empirical data provided by fabric manufacturers that captured all the nuances of spacesuit use. Fabric selection was based on intuition. As the odds of randomly selecting the suitable fabric and controlling all the variables are very poor, the fact that selections were successful with limited testing reflects a non-quantifiable talent at work. However, fabric selection was but one step toward the relative magic of spacesuits.

Every fabric has characteristics which behave differently in different directions as a result of the weaving of the fabric. This is called bias. In clothing, the direction of the bias in the fabric can cause a garment to have a form-fitting appearance and move well with the wearer or not. It can cause the garment in pressure suits to have a considerably lower ultimate capacity. This can be life critical at high altitudes or in space. Moreover, the bias in pressure suits can change the amount of effort required to move a joint. Therefore, this invisible characteristic called bias is extremely important in space pressure suits. However, before Apollo spacesuit efforts were over, people would harness the mystical art of bias to literally improve the shape, fit, and function of spacesuits.

Beyond the "soft" fabric sections, pressure suits also required hard metallic parts such as disconnects and bearings. This too had a supplier that was on almost every competitor's team. This supplier was Air-Lock.

Air-Lock started as a small family-owned machine shop located in the scenic

Figure 1.16. Gemini spacesuit circa 1965
(courtesy NASA)

shoreline community of Milford, Connecticut. At the end of World War II, David Clark started developing partial-pressure suits and needed custom, high-quality connectors, which were obtained from the Milford Tool Company in Milford, Connecticut. The Milford Tool Company had manufactured anti-G (gravity) valves and connectors for the David Clark Company during WWII, so it was a natural choice to manufacture Clark's high-altitude pressure suit hardware. In the early 1950s Clark and Bill Boynton, the proprietor of Milford Tool, formed Air-Lock Incorporated to manufacture pressure suit hardware for the David Clark Company.

Air-Lock connectors and bearings with low-effort pressure seals were critical components of high-altitude pressure suits starting in the late 1940s as well as all NASA spacesuits beginning with Project Gemini (Figure 1.16).

While Air-Lock grew over the decades, it remained loyally located in its town of origin. While the previously discussed organizations for the most part represented the aviation industry of the northeast, the west coast had a formidable technical capability. The Arrowhead Products Division of Federal-Mogul

Corporation located in Los Alamitos, California had been a pressure suit competitor starting in WWII. At the creation of NASA, Arrowhead was the main competitor of B.F. Goodrich, and these two companies were the industry leaders in high-altitude aviation pressure suits. Arrowhead was initially expected to be a major spacesuit supplier.

Litton Industries Space Sciences Laboratory in Los Angeles, California at the start of Apollo had not been previously an aviation supplier. It had been an electronics development center. Litton's "spacesuit" in 1958 was the Litton Mark I Pressure Suit that was the creation of Dr. Sigfried Hansen. Hansen was born in San Francisco and majored in electrical engineering at the University of Washington in Seattle. He married and had four sons. During and immediately after World War II, he designed radar systems before going to work for Litton Industries in the 1950s.

At Litton, Hansen took on the challenge of improving vacuum tube development. Vacuum tubes required a laborious process to assemble before performance could be tested and verified. The labor involved drawing a vacuum, potting the tube, and then letting the potting material cure. Hansen recognized that if a technician assembled the electronic portion of the tube and then tested it in a vacuum chamber, the process would be greatly accelerated. So, Hansen developed his pressure suit for this use. Regretfully, transistors soon made vacuum tubes obsolete. However, the Air Force quickly took interest in Hansen's suit giving it new life.

The Air Force helped fund evaluations of Litton Mark I (Figure 1.17) in the late 1950s. Hansen left Litton in 1959 to work for a former employer on space vacuum chambers. In NASA's Apollo survey of pressure suit technologies, NASA evaluated the Mark I, which was the beginning of an historic spacesuit development effort toward hard structure suits.

At the start of Apollo, the AiResearch Division of Garrett Corporation located in Torrance, California was a space program life support provider. In 1967 AiResearch elected to become a complete spacesuit system provider and competitor to Litton for the Apollo Block III contract. It is now part of the Honeywell Corporation, which continues to support space exploration. Litton, AiResearch, and NASA's Ames Research Center, California provided the most advanced spacesuit technologies of the Apollo era.

There was one more significant contender in the community, NASA itself. The decision to contract the development and program management of the Apollo spacesuit to a business was not universally supported within the ranks of NASA, nor was it the initial direction that NASA elected to take in lunar spacesuits. However, to understand this part of the story, one must understand the beginning of NASA.

On October 4, 1957, the Soviet Union successfully launched the world's first artificial satellite, Sputnik I, into Earth orbit. The "arms race" between the U.S. and Soviet Union instantly became a "space race" with the Soviets ahead. The United States had no rockets capable of placing the smallest of radios, let alone people, into orbit. They did not even have a national space agency to coor-

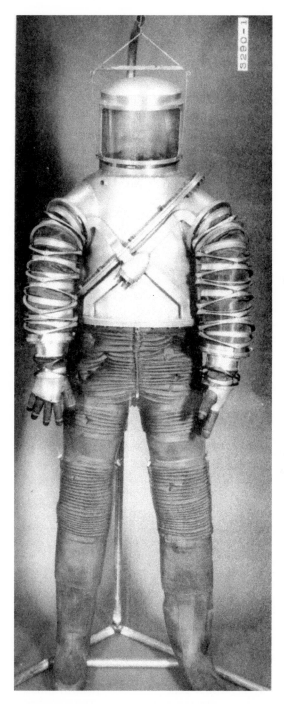

Figure 1.17. The Litton Mark I Pressure Suit
(courtesy National Air and Space Museum)

dinate and lead the effort to catch up. In less than a year, Congress created the National Aeronautics and Space Administration (NASA). The National Advisory Committee on Aeronautics (NACA), founded in 1915, was dissolved and its functions, personnel, and facilities were transferred to NASA, permitting "non-military" displays of American technical advances. NASA additionally inherited other government organizations, including the Langley Aeronautical Laboratory in Virginia, the Ames Aeronautical Laboratory near San Francisco, the Lewis Flight Propulsion Laboratory in Cleveland, Ohio, and a small high-speed flight research facility at Muroc Dry Lake in California that was the center of the nation's most advanced aircraft development. Pressure suits were an adjunct to the development of aircraft as well. It was there that NASA started its lunar spacesuit development.

The start of U.S. lunar and extravehicular space exploration efforts was actually in the summer of 1959. NASA tasked its High Speed Flight Station with creating initial requirements for spacesuits for lunar exploration in the not-too-distant future. By 1960 contractor organizations were engaged in lunar-specific efforts either by internal funding or under study contracts to NASA. This period was before the creation of mission profiles and system requirements; thus, any vision of a lunar suit system was equally promising.

Republic Aviation (Figure 1.18) and Space General Corporation both offered suit-capsule concepts that could potentially replace the pressurized enclosure of the Lunar Module and would provide the astronaut with friendlier support during all-day spacewalks.

Unfortunately, the suit-capsule concepts required additional spacecraft interior volume and added launch weight. This came when NASA struggled, in parallel, with how to create a spacecraft with enough lift capacity and interior volume to place just one person into orbit. These novel designs from the High Speed Flight Station quickly passed into history. When President Kennedy set the nation's goal to go to the Moon in 1961, NASA knew they did not have the facilities needed to support such a program. A new center in Houston resulted. Moreover, NASA had surveyed the nation's contractor community for the design and manufacturing resources required. The Houston NASA group responsible for Apollo spacesuits was the Crew Systems Division. This group started life as the Space Task Group located at Langley Research Center in Hampton, Virginia.

In September 1961 NASA's Life Systems Division, at that time located at Langley, issued contracts for pressure suit studies that included preliminary prototypes. While these 1961 studies were not neatly linked to the Gemini and Apollo programs that started in 1962, NASA's Matt Radnofsky recounted in a 1966 interview that contracts issued to Arrowhead Rubber, the David Clark Company, International Latex, and Protection Incorporated produced the first "Apollo suits."

Matt Radnofsky was a first-generation American from New Jersey who was both figuratively and literally larger than life. He had a large head, a tremendous mane of hair, and a voice which could be heard through brick and

Figure 1.18. The Republic Aviation lunar prototype
(courtesy NASA)

mortar. When he entered a room, he filled it. He was all energy and impatience, the latter showing when something or someone threatened to derail or even modify the path he was taking. If Matt was on one of his frequent quests, anyone of any stature or persuasion was seen by him as someone available to assist. He brooked no obstacles of protocol or organizational structure.

Radnofsky's impatience extended to his demanding immediate attention to any question. Even though cell phones did not exist during Apollo, NASA's primitive rotary dial devices did have an intercom feature, which simply required one to press the intercom button and dial a number. This was, to Radnofsky, too much of an inconvenience. He had nature's own intercom, his voice. He would simply sit in his office and yell for someone. One of his people, Jim Barnett, had an office across the office foyer from his. Barnett had the habit of keeping his door closed, and thus claimed that he could not hear Radnofsky call him. Radnofsky took to flinging metal paper weight letter openers, which were available by the gross from the General Services Administration catalog, at Barnett's door until he opened it and answered. Barnett never returned the openers but instead kept them forcing Radnofsky to order more.

While Radnofsky appeared to be a lion at work, he was a man of many facets depending on the environment. At home, he was easy-going, a doting father, and an obliging husband. Radnofsky was no slave to fashion. He seemed to see clothes as a social necessity but not to be taken seriously. He frequently wore a bright yellow knit tie, which sported a knot as big as a fist. His specialty had to do with fabrics and associated non-metallic materials. He had a vast network of contacts in government and industry, and he was continually sought after as the resident expert on spacesuit and survival equipment materials.

The David Clark 1961 prototype (Figure 1.19) was a derivation of David Clark's X-15 Suits. Like the X-15 model of the time, the David Clark prototype had a rear entry that was well liked in defense high-altitude pressure suits but had a zipper bulge that ran horizontally across the back when pressurized that could be uncomfortable in an on-the-back position. In comparison to the International Latex study prototype, the contour of the David Clark Gemini suit shoulders followed those of the wearer and thus took up inches less volume that would later prove to be of great benefit in the tight confines of the Apollo capsule. However, at this point in the process, the capsule volumes had not yet been established. The David Clark pressure suit could provide mobility in almost any direction with moderate effort by standards of the time.

The mobility joints of the International Latex study prototype (Figure 1.20) were easier to move but in specific directions as the convolute joints swung back and forth on their two side restraints. The movement direction of each joint was determined by where the restraints were attached on the garment.

In parallel with the suit studies, NASA's Manned Spacecraft Center was being created in Houston. This was not the Center of today but rather a scattered association of office space that occupied some 17 buildings in the greater Houston area. However, in two years an incredible transformation occurred. This had to be created in parallel with the development of the Apollo spacesuit.

At the start of the Apollo program, the site that is now Johnson Space Center was a cow pasture (Figure 1.21). Legend has it that the white cow was a NASA employee. The brown cows worked for contractors. Many in the contractor space community believe this to be true because the ratio appears correct; however, some NASA wags have countered that it shows one NASA employee was equivalent to countless contractors.

One of the young engineers who initially helped build the facility (Figure 1.22) that is now named the Johnson Space Center was Joseph J. "Joe"

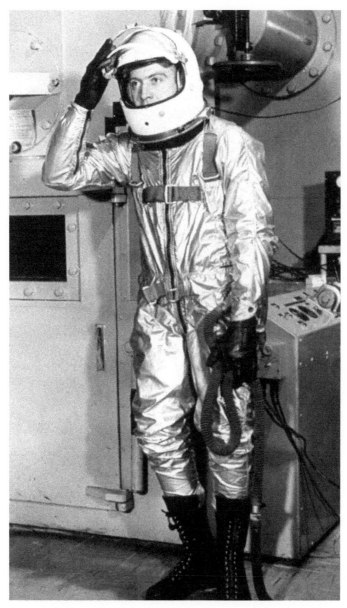

Figure 1.19. The 1961 David Clark Study Suit
(courtesy UTC Aerospace Systems)

Kosmo. Joe came from Scranton, Pennsylvania, and graduated from Pennsylvania State University initially accepting an entry-level engineering position in the just formed NASA Space Task Group at the Langley Research Center in Virginia. In January 1962 Joe transferred to Houston to help transform pasture

Figure 1.20. George Durney demonstrating the International Latex Suit
(courtesy ILC-Dover LP)*

land into the Manned Spacecraft Center needed to support the Apollo program. However, he quickly went on to help support spacesuit development, first with Mercury, then Gemini, and finally for Apollo advanced suit systems as part of the Crew Systems Division.

Joe was a man of varied interests, from archeology to strumming a guitar to spending time in a glider. On his honeymoon, Joe and his new bride went hunting for petrified shark's teeth on a North Carolina beach. This gift for curiosity and exploration was a guiding factor in his NASA career. He was one of the earliest suit test subjects for NASA, and only the fact that he had a "trick" shoulder, which could become dislocated rather easily, prevented him

Figure 1.21. The future site of the Manned Spacecraft Center in 1962
(courtesy NASA)

Figure 1.22. The Manned Spacecraft Center in 1964
(courtesy NASA)

from continuing in suited activities. His progression to advanced suit technology was only natural, considering his wide-ranging interests in human performance in spacesuits.

However, the path that produced the first spacesuits used on the Moon was somewhat different. This was also a Crew System Division Houston-based effort that started in the role of providing technical oversight and support to the Apollo suit effort. The role brought its own Apollo cast. At the top was the Division Chief, Richard S. "Dick" Johnston. His Assistant Chief was Edward L. "Ted" Hays. Under Hays was a Special Assistant for Apollo Support named James V. "Jim" Correale Jr. Organizationally under Correale, the Apollo spacesuit-related organization split into a Crew Equipment Branch that included the Pressure Garment Assembly and an Environmental Control System Branch that provided the oversight for spacecraft and spacesuit life support systems. The Branch Chief on the Pressure Garment side was Matthew Igor "Matt" Radnofsky. His Assistant Chief was Charles C. "Charlie" Lutz. The Pressure Garment Head (i.e., manager) was Jerry R. Goodman. Escape, Restraint and Support was under Douglas J. "Doug" Geier and Survival Equipment and Materials were supported by Franz Rinecker. On the life support side, the Branch Chief was Robert E. "Bob" Smylie and the Backpack Head was William C. "Bill" Kincade.

Of all the Crew Systems personnel, one who most influenced the initial years of the Apollo Space Suit Program was Matt Radnofsky. Radnofsky was one of what an associate, Joe McMann, called the "Wild Men." By this, McMann meant that Radnofsky was one of those people who were afflicted with genius, yet forced to work with more plebian folk. Others at NASA that suffered from the same difficulty were the aforementioned Jim Correale and Ted Hays. However, they made life richer by their flair, flamboyance, and technical contributions of inestimable value.

The aforementioned organizations, under the NASA umbrella, set the stage for the creation of Apollo spacesuits.

CHAPTER 2

The Initial Apollo Spacesuit Contract

Reaching the Moon before the end of the decade required almost every aspect of the Apollo program to be developed in parallel. Hundreds of systems from launch vehicles and spacecraft were competing for weight and volume allowances that would allow them to be successfully designed. Since Lunar spacesuits could have an impact on most of these parallel development activities, NASA desperately desired spacesuit development to be completed at almost the start of the overall program.

However, the unknown challenges to spacesuit development ultimately made this a parallel activity. To complicate the Apollo overall effort further, as each program struggled to meet the total weight and volume requirements to support the lunar missions, spacesuit requirements were impacted, thus increasing the challenge.

To coordinate this Herculean effort, NASA selected a systems engineering approach, which established best estimates of system specifications and control interface drawings. As long as a system did not violate the system specification or control drawing, its development could proceed in parallel with all the other systems with reasonable confidence that everything would fit and work well together in the end. Thus, the overall Apollo program from the outset had the (then) Space Suit Assembly, Command Module, and (then) Lunar Excursion Module programs operating separately with NASA having the responsibility of assuring that everything came together as a system that would successfully take men to the Moon.

The almost unobtainable art of systems engineering is establishing accurate specifications at the beginning. If the specifications are incorrect, the subsequent product will not be able to accomplish its intended purpose or not work with the other systems. If the initial requirement is incorrect, at some point that inadequacy becomes apparent. The requirements then change, and the design, interface validation, and the development process start overiteration becomes a way of life.

The beginning of the Apollo Space Suit Program was to be doubly challenged as no one knew the correct requirements or the magnitude of the technical challenges. The initial Apollo Space Suit Program between 1962 and 1965 proved an education to all. Yet, that was not really the beginning.

A PROGRAM FOUNDED IN CONFLICT

The Apollo Space Suit Assembly program began with a decision that NASA did not have the resources to be the spacesuit integrator or "prime contractor." The integrator has to oversee the development of all the components so that they fit and work well together to successfully accomplish the intended function. For Mercury and Mercury Mark II, which was renamed Gemini in 1962, NASA was the spacesuit integrator. For Apollo, NASA wanted a contractor to deliver a complete, good-working system. NASA began their process to find the right prime contractor by formulating technical requirements. NASA then issued a request for proposal. The request specified that the contract would be issued to a prime contractor who would be responsible for the delivery and performance of the entire suit system.

The prime contractor would have to have a team of expert personnel or subcontractors with all the necessary capabilities to accomplish the spacesuit development. One significant requirement was that the pressure suit had to support launch, reentry, rescue, and extravehicular activity. The contractors or contractor teams provided responses to those requirements in the form of proposals, which were due on March 30, 1962. NASA's selection of a proposal and subsequent contract negotiations would establish the initial program requirements and contractor.

The field of competing teams included the most prestigious names in the aviation and defense industries, such as the AiResearch Division of Garrett Corporation, Arrowhead Products Division of Federal-Mogul Corporation, Bendix Corporation's Eclipse-Pioneer Division of Litton Systems, B.F. Goodrich, the David Clark Company, General Electric, Grumman Aircraft, Hamilton Standard Division of United Aircraft, the International Latex Corporation, Ling-Temco-Vought, North American Aviation, Northrop Corporation's Space Laboratory, Republic Aviation, and Westinghouse Corporation.

Of the contractor teams, only International Latex is known to have produced an internally funded pressure suit prototype for the competition; it had suit system accessories to demonstrate features of the envisioned spacesuit (Figure 2.1).

The evaluations quickly came down to two teams. One was headed by a pressure suit provider, International Latex, whose team included Republic Aviation and Westinghouse Corporation. The other was led by a life support provider, Hamilton Standard, with the David Clark Company being the supporting suit subcontractor. For the competition, Hamilton performed extensive research on all the potential ways life support could be provided within the limitations of the backpack size and shape. The Hamilton proposal explained

Figure 2.1. The 1962 ILC Competition Suit with Westinghouse Backpack Mockup
(courtesy ILC-Dover LP)*

all the options with a methodical evaluation using quantitative data to support their selected backpack technology concept as being the best. Hamilton also pledged to build internally funded facilities and test rigs to support the Apollo program should they be awarded the contact. To enhance their proposal, the

Figure 2.2. Hamilton Standard commissioned 1962 painting of lunar exploration
(courtesy UTC Aerospace Systems)

parent corporation, United Aircraft, funded paintings of lunar activities with the artist being supported by their most knowledgeable engineers (Figure 2.2).

In April 1962, NASA decided to split the contractor teams into two as it desired Hamilton Standard to be the portable life support system "backpack" and overall suit system provider but wished International Latex Corporation (ILC) to be the pressure suit subsystem designer and fabricator. Hamilton had systems engineering knowledge and more government contract experience. The ILC had joint mobility which required the lowest effort, although this was limited to specific directions.

While both Hamilton and ILC pledged that they could and would work together in April, NASA's decision to split the contractor teams into two was described by many as a "shotgun wedding." In retrospect, this may have been in jest but it was certainly a clash of cultures.

In the early 1960s, the management decisions at ILC flowed from its corporate headquarters in New York City and were a blend of masterful salesmanship and tenacious negotiation that stemmed from experience in the garment industry. The headquarters were located in Manhattan's prestigious Empire State Building, which was the world's tallest building at the time. The technical talent and manufacturing capabilities, as they related to Apollo, were located in Dover, Delaware. This was frequently referred to as ILC-Dover.

The entire Hamilton organization was located in Windsor Locks, Connecticut. Its management typically rose up through the ranks of the aviation and defense business and viewed business in terms of contracts, specifications, schedules and cost, which would automatically be met unless there was a profound reason it was not possible. By Hamilton's paradigms, the negotiations between Hamilton and ILC should have taken days to reach a working agreement. In actual fact it took four months. In the meantime, NASA Houston was under pressure to provide Apollo suits to support vehicle preliminary development. To fill that need, NASA issued a direct contract to ILC for a very limited quantity of production versions of the ILC Competition Suit.

In what may have been a career-limiting act, Hamilton's planned Space Suit Program Manager, Alfred E. "Al" Reinhardt, traveled to Houston to argue the merits of using the David Clark Company as the Apollo pressure suit supplier before the start of negotiations with ILC. Reinhardt was a bright, frank, hard-driving engineer who had proven to be a highly effective manager of challenging programs. However, from Reinhardt's meeting with NASA personnel, it was clear that Hamilton being Apollo Space Suit Assembly prime contractor was contingent upon ILC being the suit subcontractor. Hamilton acquiesced. This left the people at the David Clark Company very disappointed as they felt Hamilton gave up too easily.

The negotiations between Hamilton and ILC were not simple. The negotiators came from different worlds. Hamilton's chief negotiator and planned Space Suit Program Manager was Al Reinhardt. Reinhardt was one of many project managers in an aviation industry division located across the street from a modest airport and otherwise surrounded by tobacco fields in a small town 15 miles north of Hartford, Connecticut. His office was plain with just enough room for his desk, a work table, his desk chair, and three smaller chairs. The chief negotiator for ILC was its Vice President and Treasurer, D. Irving Obrow whose spacious and tastefully decorated office was located in the Empire State Building overlooking central Manhattan.

Reinhardt was accustomed to government contract requirements and aviation industry norms. ILC had submitted an Apollo proposal to NASA defining its schedule and cost in terms of man-hour estimates and rates to accomplish the designated tasks. Reinhardt expected that the rates and schedules from the ILC competition proposal would carry over into the subcontract relationship with Hamilton. Moreover, in technically challenging or large programs the contractor usually places an on-site representative engineer at the subcontractor facility to act as an engineering and management liaison between the two companies.

Obrow brought a different experience. In his view, everything was subject to negotiation. International Latex desired higher hourly rates from Hamilton than those proposed to NASA in the Apollo competition. The ILC justification was compensation for expected lost advertising value from ILC not being the prime contractor. Hamilton's proposed placement of an engineer on-site at the ILC Specialty Products Division in Dover Delaware was absolutely unacceptable.

From April until August of 1962, Reinhardt, Obrow, and supporting personnel periodically met without progress. ILC proposed no new alternatives. Any Hamilton-suggested changes caused ILC to break off to formulate costs and schedule impact estimates, something that Hamilton always found unacceptable. Finally, the impasse was broken when Wallace O. "Wally" Heinze, the President of ILC, directly contacted William E. "Bill" Diefenderfer, the President of Hamilton. Heinze convinced Diefenderfer to grant ILC its negotiation positions. This included Republic Aviation continuing to be ILC's subcontractor to provide anatomical data developed from Republic Aviation's pressure suit programs, technical support, and pressure suit test facilities. Moreover, ILC would have direct access to NASA without having to notify Hamilton or having a Hamilton representative present.

The interaction between Heinze and Diefenderfer and the subsequent contract stipulations introduced an interesting dynamic to the process. Normally, the Program Manager of the "prime contractor" (i.e., the organization contracted to NASA) would be the program point of contact for NASA. In this case, Hamilton Standard was the prime contractor and the point of contact for NASA would be Al Reinhardt. The ILC point of contact for Hamilton would be the ILC's Program Manager, Leonard "Len" or "Lenny" Shepard. Shepard was a versatile and innovative technologist who was recruited from being a television repairman to join the ILC's Specialty Products Division to help with efforts in their helmet and later pressure suit development program, in recognition of his many talents. Although Shepard was the ILC Pressure Garment Assembly Program Manager and faithfully attended virtually all the meetings, during the first two years of the program his name would rarely appear in documents addressing program issues. In the late 1960s, he would distinguish himself as the leader of ILC's pressure suit efforts, Any key decisions or disagreements appear to have been elevated to Diefenderfer and Heinze for resolution.

In September 1962 Hamilton was allowed by NASA to direct ILC to proceed in advance of the formal contract award. The contract to Hamilton followed in October 1962.

Even before the contract was formally awarded, the Apollo spacesuit challenge was increasing. The announcement of the winners in April 1962 was accompanied by a 6% increase in the required average life support capacity. With the formal award of the contract, the average life support requirement was increased again. This time it was increased by 86%, which was almost twice the original rate. Also at this point, a maximum hourly metabolic energy expenditure rate was added, which was over three times NASA's original requirement. This meant having to provide over three times the oxygen, carbon dioxide removal, and cooling capacity. All the while, the volume, weight, and development schedule allowances for the backpack did not increase.

Beyond the spacesuit assembly and backpack levels, Apollo Pressure Garment Assembly development also had technical challenges. Apollo was the first U.S. space program to take a promising pressure suit prototype and develop it into a functional suit system that would reliably and effectively meet a space application. Lacking past experience, no one in the process knew the magnitude of the challenges that were ahead. Adding to the difficulties was the fact that there were no clear and quantifiable suit mobility requirements other than being able to rise from an on-the-back position after a fall. The pressure suit mobility requirements in the first year of Apollo were principally being defined as adequate to meet the mission requirements in the judgment of NASA evaluators. The detailed mission activities, which were the real drivers for requirements definition, had yet to be developed. Another development barrier was the subjective judgment of "adequate" or "inadequate" that could vary significantly from one suit subject to the next.

Hamilton had recognized that the initial measure of space pressure suit quality meant extensive human testing. In preparation for the Apollo Space Suit Assembly competition in late 1961, Hamilton selected a small internal group of engineers to become pressure suit knowledgeable. This training included limited suit evaluation experience. As part of the ramp-up after the October 1962 formal contract award, Hamilton had to expand that knowledge base to create the infrastructure to evaluate spacesuit mobility and successfully certify all structural and environmental requirements. This caused the creation of "Hamiltonauts" who had to meet NASA astronaut anthropomorphic and physical condition requirements. One of the Hamiltonauts was Edgar H. "Ed" Brisson. Brisson earned his Bachelor of Science degree in Mechanical Engineering from the University of Hartford and joined Hamilton as a development engineer in 1962. Like the other Hamiltonauts, Brisson was expected to run and work out to become and stay the physical equivalent of an astronaut to provide the most meaningful suit testing possible during development.

There were also organizational challenges. At contract award, neither Hamilton nor ILC had space business units or staffed spacesuit programs. Both were expected to rapidly build from a very limited staff to being fully staffed with functional programs to rapidly meet their respective contractual obligations that were to span less than one year.

Before either organization could be free to start performing their specialties for the Apollo spacesuit, they had to establish how the backpack would attach to the pressure garment and where the controls would be located, as these features dictated the internal and external designs of both the pressure garment and the life support backpack. This brought into contact three people who would be key to the development of the first Apollo Space Suit System. These were Hamilton's Andy Hoffman and Earl Bahl, and ILC's George Durney.

Andy Hoffman graduated from the University of Connecticut and joined Hamilton in 1953. Short in stature, large in personality and universally liked, his talent for being very amiable while staying focused on the task probably resulted in his selection as Hamilton's backpack Project Engineer/Manager. His appearance and demeanor were misleading. His voice was gentle and his manner always respectful and controlled. However, one soon learned that his gentle voice belied an iron reserve and his gentle manner was relentless in tireless pursuit of excellence. Hoffman's mission was to produce man-rated examples of the primary life support backpack and the backup Emergency Oxygen System, called the EOS, within 10 months. This translated into working days, evenings, and weekends. Fortunately for him, he had and still has an understanding wife.

To establish the location of the backpack controls and how the pressure garment and backpack would attach, Hamilton Mechanical Design Engineer Earl Bahl designed a facsimile of the then-current backpack concept (Figure 2.3) to allow evaluation by International Latex (Figure 2.4). Bahl graduated from Pennsylvania State University and joined Hamilton in 1959. When Bahl joined Hoffman's backpack team, he was single and could work long hours without adversely affecting his home life. However, that would change. During the development of the first Apollo backpack, Bahl got engaged. Because the backpack efforts required he work around the clock, his fiancé was concerned that he might not be able to get time off to get married. Fortunately, accommodation was made. Bahl was exceptional in his technical capability, but another attribute which made him extremely valuable was his dedication to the truth: to tell it "like it is."

The backpack model designed by Earl Bahl, as was the norm for modeling at that time, was made of wood that was filled, sanded, and painted to look both realistic and attractive. The backpack was hollow to minimize weight. The backpack facsimile was then hand-couriered to ILC where George Durney evaluated it wearing an SPD-143 suit. Durney studied at the University level but did not complete his degree in engineering; however, he had a talent for seeing and understanding technical issues and finding solutions that worked. The backpack test started with Hoffman strapping the mockup to Durney who was already in an unpressurized suit. When the suit became pressurized, the mockup blew apart. Durney and Shepard laughed. Shepard said "well back to the drawing board." Hoffman returned to Windsor Locks with a box of debris. For the next couple of weeks, Hamilton's people worked to recreate the mockup. Instead of wood screws to hold strap brackets, the mockup was assembled with thread steel rods running from front to back. The harness was remanufactured with a belt material with imbedded steel fiber. Additionally, the interior cavity of the mockup was filled with lead to bring the backpack to the correct weight. It was then carefully repainted to conceal the previous damage.

When Hofmann arrived at ILC to redo the test, he strapped the pack up tight. The ILC personnel clearly noticed but said nothing. The suit was then pressurized. The backpack held fast while the suit deformed. Durney appeared to be uncomfortable but did not stop the test or acknowledge his condition. The evaluation went on for six hours to check suit-to-life support interfaces and mobility to accomplish control adjustments on the life support system. Hoffman was in no hurry and was enjoying the fit check. Durney wanted to

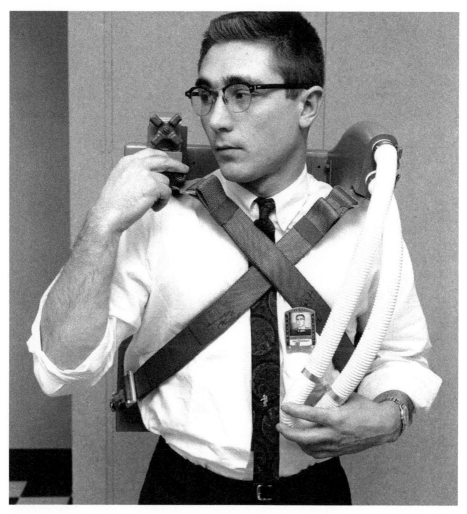

Figure 2.3. Hamilton Standard's Earl Bahl and the 1962 Backpack Development Fixture
(courtesy UTC Aerospace Systems)

quit, but Hoffman insisted the evaluation continue to gain a better understanding of the potential design issues. After this, technical-level dialogs between Windsor Locks and Dover personnel, unlike their management-level counterpart's communications, tended to be more accommodating and productive.

In Apollo, Hamilton and ILC were not the only spacesuit contractors in conflict. B.F. Goodrich tried to stem the migration of talent and knowledge to International Latex for Apollo. To that end, Goodrich resorted to legal action in "B.F. Goodrich versus Wohlgemuth." Donald W. Wohlgemuth graduated from the University of Michigan in 1954 with a Bachelor of Science degree in chemistry. Soon thereafter he was hired by B.F. Goodrich. Following a short

Figure 2.4. ILC's George Durney using the Backpack Test Fixture
(courtesy ILC-Dover LP)*

period of service in the U.S. Army, he returned to the Goodrich Company in 1956 and was assigned to work in the pressure spacesuit department. As his technical knowledge increased, Wohlgemuth was promoted to successively more important positions such as Materials Engineer, Product Engineer, Sales

Engineer, Technical Manager, and finally Department Manager. As Manager, he was responsible for all engineering of pressure suits and ancillary equipment, both in development and production phases.

In November 1962 Wohlgemuth elected to leave for a position with International Latex. Goodrich immediately filed an injunction. The Ohio court ruled in favor of Goodrich in May 1963. Although it was contested, the protracted legal action successfully precluded Wohlgemuth from ever working on the Apollo suit program.

Given the challenges that were experienced with ILC pressure suits in the first two years of Apollo and the program breakthroughs in suit shoulder width and upper-torso mobility which came from B.F. Goodrich in late 1964, it is interesting to speculate how history might have been different if Wohlgemuth, a man of obvious talent and experience, had been allowed to work on the program.

THE ORIGINAL APOLLO PRESSURE GARMENT ASSEMBLY CONTRACT

In its contract to NASA, Hamilton pledged to produce a complete, certifiable Apollo Space Suit Assembly in ten months. In its contract to Hamilton, ILC agreed to a nine-month, three-phase Apollo Pressure Garment Assembly (aka suit) effort. In Phase A, ILC would produce two prototypes to develop torso mobility by April 21, 1963. Phase B was to produce one prototype to initially develop thermal protection by May 7, 1963. By July 21, 1963, ILC was to produce two more prototypes to refine and complete thermal protection development.

In parallel with suit development, ILC also agreed to an eight-month contract for an initial training fleet of 20 pressure garment assemblies (Figure 2.5). These first training suits were essentially production copies of the ILC competition prototype suit that NASA designated the AX1L. Except for the never-ending bickering over the name of these suits and for quality issues relating to the first unit that took two and a half years to resolve, the interactions between Hamilton and ILC went relatively smoothly and training suit production was on time with good cost performance. The name dispute was that ILC wanted to call them AX1L suits. As these first training suits were different from the ILC Competition Suit because of NASA-requested changes, Hamilton wanted to call them AX1H, to reflect the design was during the Hamilton-Apollo contract. ILC found that unacceptable as there was no Hamilton content in the pressure suit. A "compromise" was found in which Hamilton directed ILC to identify them as SPD-143, the ILC internal model designation for that configuration. Hamilton directed that there should be no reference to AX1L on the suits or delivery documents. However, on every suit label and in every suit delivery document, ILC put both SPD-143 and AX1L.

The Pressure Garment portion of the Apollo program did not follow the government contracting norm. Jerry Goodman was NASA's Apollo Pressure

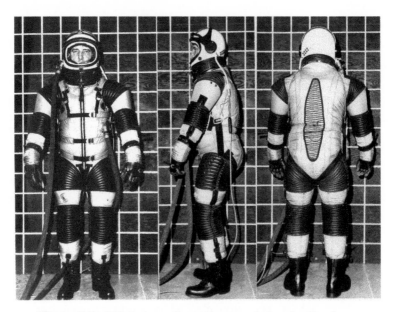

Figure 2.5. Old-design suit sporting new-design Apollo gloves
(courtesy ILC-Dover LP)*

Garment Manager. Born in Jacksonville, Illinois, he graduated from Purdue in 1958. The timing of his degree led him to become Structural Test Engineer at McDonnell Aircraft testing Mercury and Mercury Mark II (later renamed Gemini) spacecraft parts. This led to spacesuit testing at Wright-Patterson Air Force Base in Dayton, Ohio and being invited to head this key part of the Apollo spacesuit effort. One might expect that Goodman was the most influential person in the Apollo pressure suit effort or perhaps second after Al Reinhardt who was Hamilton's Program Manager. However, the historical trail shows they were overshadowed by Matt Radnofsky, Goodman's boss, and Wally Heinze, the President of ILC.

The ILC-Hamilton suit development contract did not progress far before running into difficulties. In one of the first meetings between Hamilton and ILC to work out details going forward, Al Reinhardt lost his patience and yelled at Obrow. Obrow was incensed. Heinze ordered no further meetings with Hamilton. Diefenderfer tried but was unable to placate ILC. By mutual Hamilton/ILC agreement, Roger D. Weatherbee replaced Reinhardt as Hamilton Space Suit Assembly Manager. Weatherbee was an equally bright engineer-manager, but brought a more reserved, polished, and gentlemanly style. ILC pledged to Hamilton that all contract requirements would be met. Hamilton in return agreed to provide minimum oversight. These agreements started a short-lived period of peace and cooperation between the two companies.

The first new Apollo pressure garment development proved to be gloves. In the spring of 1963, ILC introduced the new Apollo glove design into training suit production (see Figure 2.8).

The next new area of Apollo suit development also came during the production of the SPD-143. This was the beginning of thermal overgarment prototyping and evaluations. In the first five years of the Apollo program, the approach to thermal and micrometeoroid protection was the use of separate overgarments worn over the pressure suit assembly before leaving the spacecraft.

The first concepts were for donning a single coverall like a skimobile suit. Later, it was found that separate garments like ski pants and a parka coat were easier to don. Thermal outer-garment development started with the creation of mockups for donning evaluation and then progressed to prototype units that could be evaluated for actual thermal protection capabilities. In this initial attempt, ILC exclusively performed the development. The garments were first tested at Hamilton, and then forwarded to NASA for further evaluation. The garments were found to be inadequate for both thermal and particle impingement protection. As a result, Hamilton assumed the engineering lead for the overgarments. The next generation of prototype overgarments tested in 1964 were many times thicker.

The first suit development issues were not performance related. They were concerned with cost and schedule. By April 1963 two prototypes were expected to be delivered, thus completing mobility development. April came and went without any prototypes. In May 1963 ILC informed Hamilton of an expected overrun equal to 44% of their development contract. Without suits to test, cost and schedule rocketed to the top of management's attention both at Hamilton and at NASA. In retrospect, this probably should not have been a surprise. While ILC was a major corporation, its Specialties Product Division had produced only infrequent pressure suit prototypes for the Air Force and NASA prior to the Apollo contract. Simultaneously committing to production of 20 training suits in eight months, plus developing new Apollo gloves, helmets, torso assemblies, thermal garments, and overgarments while getting used to NASA contracting and quality requirements was a monumental challenge.

In July 1963 two first-design new-build prototype suits arrived at Hamilton. These deliveries made the definition of a "delivery" a topic of discussion. Torso mobility development had started with section-by-section modification of an existing training suit. ILC replaced a mobility element one side at a time (Figure 2.6) allowing the new feature to be comparatively tested against the unmodified side that represented the pre-competition mobility technology. George Durney personally did the testing. This made ILC confident that significant mobility improvements had been accomplished. An identical first-design new-build suit was then built. This development process highlighted an issue: Could the first prototype made from an already delivered training suit be counted as a "new-suit" delivery? ILC felt the answer was yes. Hamilton and the NASA Contracting Officer disagreed. How to count this as a contract item would take months to resolve.

In the Gemini and Apollo programs, the assigning of NASA designations to suit configurations was not a formal configuration management system directed

Figure 2.6. Incremental development of the first new Apollo design
(courtesy ILC-Dover LP)*

by NASA headquarters. Rather, it was an informal function provided by a quiet, thoughtful, well-respected NASA engineer named Charles C. Lutz who provided a trail to what otherwise would have been historical chaos by assigning model designations that reflected program, manufacturer and where each model fell in the development process.

Charlie Lutz was a smallish, gently rounded man who came to NASA from Wright-Patterson Air Force Base. He was non-degreed, but had vast experience in aircraft pressure suits. He was noted for not overburdening the listener with needless talk. When he did speak, it was usually after a lot of listening, and with much eye contact to assure himself that he was being heard. One NASA manager, when asked how he was going to conduct himself in an upcoming meeting which threatened to be contentious, replied "I'm going to sit next to Charlie Lutz, and say a little less than he does." His legacy lives on in his son, Glenn C. Lutz, who is a highly respected manager at NASA's Johnson Space Center.

One of the Apollo issues partially addressed by the first Apollo design was eye protection. Without Earth's atmosphere acting as a filter, sunlight in space is so intense it literally can be permanently blinding. For the first-design deliveries, ILC had produced a solid aluminum replication of a sun visor. In parallel with ILC's developments, Hamilton put Perkin-Elmer of Norwalk, Connecticut under contract for the successful development of the Apollo sun visor gold-coating process. In the late 1960s NASA would elect to move the process facilities to have it performed at Goddard Space Flight Center.

The new prototypes were first tested at Hamilton. At that point in the program, there were no quantitative ways of measuring mobility element resistance to movement or quantitative requirements for mobility. "Reasonable effort" to move or hold position was the principal criterion. Needing a means to quantify Apollo mobility improvement, Hamilton comparatively tested the new-build, first-design suit against an existing training suit (Figure 2.7).

Figure 2.7. Comparative testing of new and old designs
(courtesy UTC Aerospace Systems)

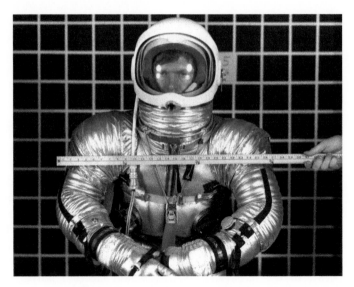

Figure 2.8. Measuring shoulder width
(courtesy UTC Aerospace Systems).

Mobility progress was questioned by Hamilton. Given the method ILC used in development, ILC's personnel had difficulty understanding Hamilton's findings. Since effort was subjective, Hamilton looked for a means of establishing what was and was not acceptable, and took pictures of the bruising incurred by the suit subject. The subject was Ed Brisson. Hamilton provided photographs of the bruising to ILC, but resolution of this issue eventually required NASA participation, which came months later. This was but the beginning of a phenomenon in which the performance of ILC test subjects exceeded that of flight crews and customers. ILC suit subjects spent so much more time in the suits that they had, probably unknowingly, mastered techniques such as speed and direction to get optimum mobility with less effort and less consequent bruising. However, understanding this dynamic would take many years.

Hamilton's evaluation further found the shoulder width was greater than the contract requirement (Figure 2.8) (i.e., too wide). Otherwise, the suit was deemed acceptable. ILC assumed there would be additional Command Module volume available and anticipated the shoulder width allowance to change. However, it appears that they neglected to inform Hamilton in advance that they were adhering to the shoulder width requirement.

The AX1H-021 suit was then sent for its NASA debut where it experienced an entry zipper failure and decompressed. The suit was sent to ILC, repaired, and returned to NASA for evaluation during which NASA objected to the stance of the suit when pressurized (Figure 2.9) but not the mobility or shoulder width. While this left the need for further development in question, Hamilton directed that stance, mobility improvement, and shoulder width compliance development should continue into the AX2H configuration.

Figure 2.9. Ed Brisson demonstrating a first-design suit
(courtesy UTC Aerospace Systems)

Even if Hamilton and NASA had agreed on the evaluation results and ILC's New York headquarters had accepted that there were issues with the Apollo pressure garment designs, the amount of time required to make such changes was greater than that available. The remaining schedule was compressed as a result of delivery slippage. Even the original eight-month schedule was too short for three design iterations each of which required testing to develop feedback and then have the feedback incorporated into the next design iteration. The race to the Moon had framed the scheduling of this design effort such that success was the only acceptable outcome. This created even greater pressure on all participants to assume there were no issues with the design.

This first Apollo suit design also included a Russel Colley creation. Apollo was proceeding in parallel with NASA's ongoing Mercury program. The

Mercury program recognized that there could be situations where an astronaut could be sealed in a pressurized suit for long periods of time and need the ability to drink. Colley came up with a solution by inventing the Helmet Drink Port (Patent No. 3,067,425). This allowed an external bottle with a mechanical "straw" valve device to be inserted into the helmet (lower center in Figure 2.6). While helmet designs changed frequently in the early years of the Apollo program, the drink port continued to be a helmet feature through the lunar and Skylab missions.

In early September 1963 the first of two second-design pressure suits was completed. This was the AX2H-023 prototype. This suit was delivered in October, and was built as a prototype for a second fleet of training suits. The prototype featured white nylon outer fabrics to facilitate cleaning and reduce manufacturing cost. The stance of the first-design suit had been corrected. The second-design suit had the same base mobility architecture as the first, but the shoulder width had been reduced, although still not within specifications. Reduction of the shoulder width had resulted in a slight reduction of shoulder mobility; however, some additional lower-arm mobility had been gained.

While the management of ILC and elements within NASA had not yet accepted mobility as an issue, ILC was proactively responding to cost in its development efforts. This was probably in anticipation of the forthcoming contract for 38 new-configuration training suits. ILC internally funded a prototype where the life support system (LSS) connections were in umbilicals attached to the front of the torso by short hoses. The LSS connectors were inexpensive, off-the-shelf, commercial units. Since the Apollo program LSS connectors were custom-designed for the Apollo program and highly expensive, this had significant potential to reduce program costs. This also had the potential to make the donning and doffing process for EVA much easier as the connections would easily be in sight. Demonstration of this prototype to NASA resulted in NASA directing Hamilton that the next design iteration, the AX3H, had to have these features.

Pressure suit mobility rose to be a significant issue in the eyes of NASA in November 1963 with the test of a second-design suit in a reduced lunar gravity simulation. This was one of the few definitive mobility requirements in the original Apollo spacesuit contract. Equipped with a volumetric representation of the then-current portable life support system shape, subjects were unable to rise from an on-the-back position. Additionally, NASA found the downward visibility of both the first-design and second-design helmets to be unacceptable. NASA directed a helmet redesign to correct this condition.

The inability to rise from an on-the-back position in an Apollo suit caused NASA to find Hamilton to be in failure of meeting contract requirements, which is serious in the aerospace industry. Additionally, Hamilton was criticized for not adequately managing ILC. Hamilton was required to promptly provide corrective action. What followed may have been an unfortunate failure to adequately communicate as a result of cultural differences.

The NASA notification was telegraphed to Hamilton. Bill Diefenderfer,

Hamilton's President, immediately drafted a company letter requesting ILC to provide proposed corrective actions within 48 hours. A courier then drove to New York City to hand-deliver the letter to ILC's corporate headquarters, which was the method of communications throughout the Hamilton-ILC Apollo program. It appears the courier was asked to wait while Wally Heinze drafted a reply. The Heinze letter was then driven back to Windsor Locks, Connecticut. The letter stated that ILC was aware of the testing and the problem was due to lack of NASA suit evaluator training. There were no mobility issues with the suit, thus no corrective actions were required. To Diefenderfer, this was certainly taken as arrogance and defiance on the part of ILC. However, this may not have been the case.

It may have been that Heinze was operating from a garment industry mindset. In the garment industry a supplier would never willingly take responsibility for a problem and offer up corrective actions as such actions would be at the supplier's expense. However, this was not the case in government contracting. The contractor almost always gets reimbursed for all expenses and if the problem could be argued as being unforeseeable, as space exploration was a new field, the contractor would probably get a fee (i.e., profits) on the activities plus more contract schedule. The ILC personnel in Dover would have funded-in an expanded opportunity to demonstrate their technical talents and provide the solutions before there was a program crisis with their customer looking in other directions for help. Clearly, this could have worked to ILC's advantage and profoundly changed subsequent history.

From a government contracting perspective, ILC's reply did not make sense. First, denying a problem is not a viable option as the government contracting agency would suspend the contractor's ability to make deliveries and get paid by the government. Filing a challenge on the basis that the government was wrong would not be much better. The burden of proof would be on the supplier. Proving the government wrong to the satisfaction of the government auditor is unlikely and the subsequent corrective actions would certainly be much more painful to satisfy. Consequently, Heinze's reply appears to have been poorly considered. Heinze probably thought Diefenderfer would write back in a day or so and the dialog would continue. That would not be the case.

Diefenderfer had 72 hours from receiving the telegram to provide corrective actions, or give an explanation of why such actions could not be provided in time, accompanied with a schedule of when the action plan would be delivered. Diefenderfer clearly took Heinze's letter to mean that ILC was not going to participate in creating corrective actions and tasked his Space Systems personnel with creating a plan that would appease NASA and not require willing ILC support.

The actions Hamilton offered to NASA included ILC continuing mobility development in a third design effort, Hamilton supplementing the ILC suit design effort with Hamilton-internal resources, Hamilton assuming the helmet redesign so ILC could concentrate on torso redesign activities, and Hamilton developing quantifiable/measurable mobility requirements.

While this pacified NASA, it had severe negative impacts on the already difficult working relationship between Hamilton and ILC. ILC's Specialty Products Division was strongly identified with helmets as it had gained its entrance into high-altitude aviation suits through helmets with masks, and the production of partial- and full-pressure suit helmets. With this latest development, the pattern of presidential correspondence between Bill Diefenderfer and Wally Heinze changed. The "Bill and Wally" dialog would no longer be the venue for resolution of disagreements between Hamilton and ILC. The paramount program issue as far as Heinze was concerned was ILC continuing to be the helmet provider. He was not ready to accept that there were shoulder width or mobility issues. For Diefenderfer, shoulder width and mobility required immediate attention. These organizational leaders were resolute in their different positions. In the coming year, program decisions would result in a clash between Diefenderfer and NASA's Matt Radnofsky. This had technical and programmatic consequences.

The corrective actions set Hamilton on a path toward pressure garment design and manufacture by pledging engineering support to ILC's mobility development efforts and by assuming the design and manufacture of a planned new helmet. In parallel with ILC's third design effort, which produced the AX3H suit, Hamilton formed a very small specialized engineering group under the leadership of Mark Baker in November 1963. The group included John Korabowski who would be a key contributor to Hamilton's later efforts.

Korabowski was a large man, six feet tall and broad through the shoulders and hips. With slicked back hair, his large round face clearly showed his good nature. He was always friendly and positive. Not having the opportunity to go to college, he was nevertheless very knowledgeable in mechanical design and manufacturing engineering. He knew exactly what machine and process to use to produce any part to meet a drawing's requirements. His voice was very distinctive. It was higher pitched than one would think for a man of his size but it could have a little growl to it when it gained body and volume. He spoke very loudly as he was accustomed to talking over the noise of the lab or factory floor.

The first product as a result of this Hamilton effort was the "Multi-Directional Suit Joint" elbow. The group then developed a tapered bellows. Both designs were based on the convolute molded by ILC-Durney, but used a complex system of cables, pulleys, and rings instead to gain a multidirectional ability. This system showed sufficient promise that NASA provided a 1963 training suit for Hamilton to retrofit with Hamilton shoulders and arms. The result was the first Hamilton "Play Suit," which was tested along with Apollo and Gemini suits in Command Module evaluations conducted at North American Aerospace in March 1964. The Play Suit received unfavorable reviews for comfort and showed only minor improvements in mobility over the ILC Apollo suits. However, it gave Baker and Korabowski greater insight into pressure suits that would influence later developments.

During 1963 astronauts and other evaluators with experience of both the

ILC Apollo and David Clark Gemini suits expressed a preference for the David Clark rear-entry system. In November 1963 ILC proactively approached Hamilton requesting program funding to explore rear-entry systems. Hamilton declined, wanting all resources to be focused on torso mobility development instead.

In parallel with the second-design testing, NASA issued a request for proposal or "RFP" for 38 training suits based on the second design. Hamilton being expected to submit the proposal to NASA on the day after Christmas is a reflection of the no-time-off nature of the Apollo program. As a result of the aforementioned findings regarding the AX2H design, NASA cancelled the RFP in January 1964.

At this point, Hamilton learned that ILC had submitted a competing proposal for a direct-to-NASA contract. Hamilton felt that this was a violation of its contracts with the ILC and NASA. NASA's Matt Radnofsky and Charles Lutz were in support of a direct contract with ILC based on expected cost savings. The cost savings appeared significant. ILC quoted slightly lower hourly rates to NASA than they had with Hamilton. Hamilton's proposal involved increasing the ILC quote based on program cost overruns experienced to date and having to provide oversight. ILC's direct proposal expected no cost overruns.

Hamilton subsequently met with Dick Johnston, the Chief of the Crew Systems Division and Radnofsky's superior, to present why Hamilton should remain the Apollo pressure suit sole source. Johnston concurred with Hamilton and on January 29, 1964 issued a program memorandum to Maxime A. "Max" Faget, the Assistant Director for Engineering and Development at the Manned Spacecraft Center in Houston providing sole source justification. However, this was not the final resolution.

Unwilling to take no for an answer, Radnofsky responded the next day with a rebuttal memo. In a parallel move, NASA reduced the number of suits of the new training fleet from 38 to 14 as a result of mobility development issues, thus reducing potential cost savings. Both Johnston and Radnofsky remained firm. On February 10, 1964, a second program memorandum initialed by Ted Hays and signed by Johnston reaffirmed the single source decision probably in the expectation that this would bring closure to the issue.

While further details of this internal NASA conflict have been lost, a partial compromise appears to have resulted. By February 10, 1964 the first suits in the disputed contract were probably already under construction. Perhaps as a result, NASA issued a contract to ILC in time to support the March delivery of three ILC Apollo suits directly from ILC. These were essentially production versions of the latest suits delivered to Hamilton. These were most likely considered necessary to the program to support a Human Engineering Criteria Mobility Analysis Review or "HECMAR" being performed at North American Aerospace in conjunction with the Command Module Program. This testing was being conducted in parallel with Lunar Excursion Module Program testing at Grumman. NASA subsequently issued a contract to Hamilton for 27

Figure 2.10. The third new design, original configuration
(courtesy UTC Aerospace Systems)

training suits based on the expected third design of the Hamilton Space Suit Assembly Program. This number would soon be significantly reduced. Moreover, the total number of Hamilton contract prototype deliveries was formally reduced from five to four in light of the first-design prototype being created by incrementally retrofitting an existing training suit.

On February 20, 1964 the third-design prototype suit (AX3H-024, Figure 2.10) was delivered to Hamilton for testing. The third-design shoulders and arms were refinements of the previous designs and still did not meet the Apollo shoulder width requirement. Hamilton evaluations indicated that mobility had only marginally improved and was not yet adequate.

Concurrent with this third design, ILC produced an internally funded state-of-the-art or "SOA" Command Module Pilot suit. With this effort, ILC not only created a new prototype, it also introduced a new way of thinking about suit architecture within the Apollo program. In the Apollo program in 1962-63 the mindset had been that the Lunar Excursion Module crew extravehicular or "EV" suits and the Command Module Pilot or "CMP" suit would all have the same features and volumetric requirements. ILC's recognition of the potential advantages of having more specialized EV and CMP suits caused ILC to fund this new prototype. The premise for the SOA-CMP was that it could have narrower shoulders, with modestly less mobility, which would be less of an impediment to the pressurized suit operations of three suited astronauts within the Command Module. ILC delivered the SOA-CMP suit with the first Apollo A-2L direct-to-NASA contract suits for evaluation in HECMAR.

Command and Lunar Module evaluations in March 1964 brought more attention on Apollo pressure suit development and delayed the start of the then-planned third-design training suit production. The Command Module evaluation included three David Clark Company Gemini suits (Figure 1.16). The evaluation confirmed the Apollo suit shoulder widths were a major problem and concluded that the David Clark Gemini suits were the only ones adequate for Command Module operations. Command Module evaluations showed ILC Apollo suit durability was also problematic.

Lunar Module evaluations conducted at Grumman included the third-design prototype of the Hamilton Apollo program, the AX3H-024 suit. NASA astronauts were also part of the evaluation. In this evaluation, the suit soon experienced a failure of the ILC design umbilical-to-suit interface. The repaired umbilical attachment failed a second time causing the suit to decompress while under test by a Grumman evaluator. The astronaut ratings were unanimously unfavorable. To show his dissatisfaction with this latest ILC prototype, Astronaut Gordon Cooper elected to announce to an assembly that included top management of the Space Suit and Lunar Module programs, important dignitaries from NASA's Washington headquarters, and hundreds of onlookers that "I would not go to the moon in that suit." He additionally added that he much preferred his Gemini suit. This sent shock waves through the Hamilton Space Suit Program and created a two-fold crisis for NASA, as the new-configuration Apollo training suits that NASA desperately needed were again stopped and NASA had no requirements against which to create an adequate new suit design.

THE FIRST APOLLO PORTABLE LIFE SUPPORT SYSTEM "BACKPACK"

In late 1962, an early training suit was sent to Republic Aviation for manned testing (Figure 2.11). The test results raised concerns about the human effort required to do lunar exploration. This was dismissed at the time because the forthcoming new designs for Apollo spacesuits were expected to have lower effort mobility and improved ventilation systems. However, Hamilton took this as a strong indication that the requirements for Apollo spacesuit life support

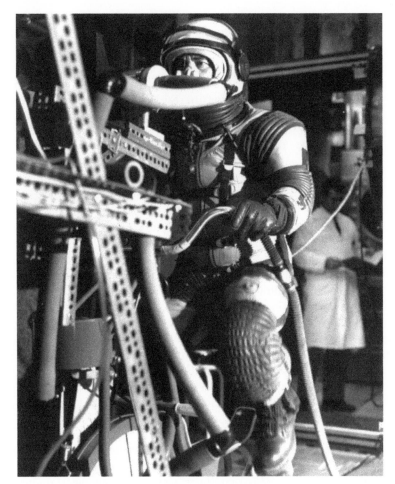

Figure 2.11. State-of-the-art test facilities in 1963
(courtesy UTC Aerospace Systems)

capacity might significantly increase and initiated internally funded research and development in "backpack" technologies.

For Apollo to be a success, the spacesuits and the vehicles had to operate as one system. To evaluate the compatibility of parallel Lunar Excursion Module and spacesuit development, manned evaluations were required. Early Apollo suits were also used for these first evaluations. To allow realistic testing, Hamilton was tasked with providing a fully autonomous (no umbilicals) ground evaluation Apollo backpack. It featured a fiberglass shell, which represented the then-largest allowable volume, and a compressed air bottle to supply a purge flow to remove carbon dioxide and humidity while providing modest cooling sufficient for adequate comfort during testing. The autonomous nature of this suit-system allowed the manned Lunar Module evaluations at Grumman

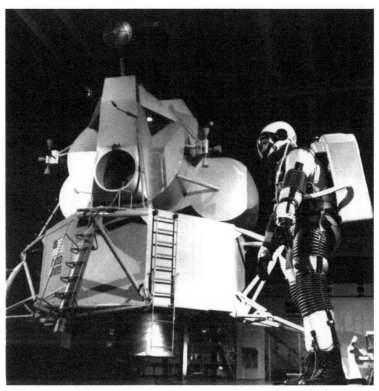

Figure 2.12. Early suit testing at Grumman
(courtesy UTC Aerospace Systems)

to be conducted in early 1963 (Figure 2.12). This unit was also used in conjunction with prototype thermal overgarments to aid suit system development. These evaluations led to the backpack shape and size being changed three times in the first six months. The changes were driven by Lunar Module hatch sizes and internal clearances.

In parallel with development of the first new-design pressure suits, Andy Hoffman's group developed the first Apollo backpack (Figure 2.13). This had progressed from not knowing how the life support might attach to a fully developed backpack certified safe for manned use in a grueling nine and a half months. The system used gas cooling, like air conditioning for a home, office, or automobile, for heat removal and was capable of sustaining an astronaut according to the 1963 requirements for at least four hours of lunar exploration. The system contained an oxygen tank, an oxygen regulator, a fan, a lithium hydroxide (LiOH) canister for carbon dioxide removal, an elbow-shaped water separator, a wick-type water boiler with a temperature control valve, and a battery. In this system the ventilation gas carried the heat from the suit and areas within the backpack to the water boiler, which evaporated water to the vacuum of space to remove heat from the spacesuit.

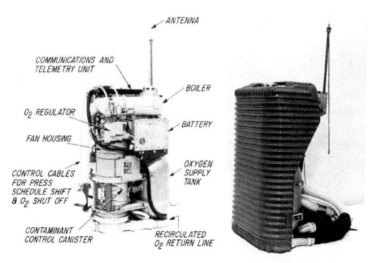

Figure 2.13. The 1963 "Gas Cooled" Apollo Backpack
(courtesy UTC Aerospace Systems)

Manned testing was conducted using a first-generation Apollo prototype pressure suit. The backpack was found to meet the then-current contract performance requirements. This would have been a great success but for manned testing (Figure 2.14) confirming that the 1963 life support requirements were not sufficient to meet the needs of lunar surface missions. More importantly, the astronauts would experience unacceptable levels of dehydration because heavy sweating was required for the removal of body heat. These findings made the first model of the Apollo backpack immediately obsolete.

The metabolic rate capacity requirements for the Apollo suit system increased again in 1964 according to the final Apollo spacesuit specifications, which stipulated an increase of 29% under normal operation and 25% under maximum use over the previous requirements iteration. Again, the volume and weight constraints for the Apollo backpack did not change. This final increase required invention and development that was beyond the life support technologies available in 1963. Lunar exploration in a spacesuit literally involved more work than had been previously envisioned. Astronauts drinking large volumes of water and perspiring profusely were not options for keeping cool and avoiding dehydration or heatstroke. "Turning up the air conditioning" inside the spacesuit was also unacceptable as the noise inside the suit from the gas being forced through the ventilation system would have been intolerable. Probably no one knew it at the time but manned demonstration of a satisfactory system would take another two years.

As part of the suit's life support effort, Hoffman also led the development of the first Apollo Emergency Oxygen System or "EOS." This was to provide backup life support. There were actually two EOS systems developed by

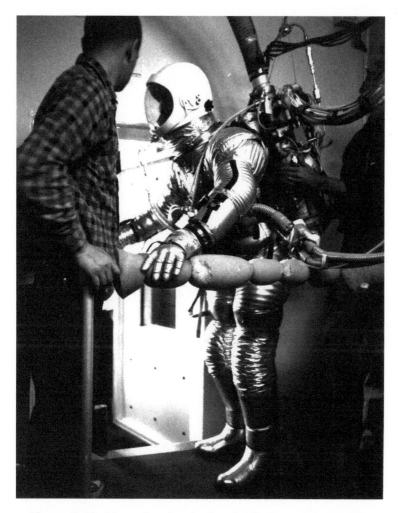

Figure 2.14. Manned testing the first Apollo Space Suit Assembly
(courtesy UTC Aerospace Systems)

Hoffman and his supporters for Apollo. The first was completed by August 1963. This was to be mounted on the backpack to provide five minutes of backup life support. Despite this first Apollo EOS also meeting all the program's requirements, the EOS would be completely redesigned to reduce weight and volume to aid the Apollo Space Suit Program in meeting the backpack's increased life support requirements.

At the point when initial Apollo Space Suit System development had been expected to be completed, only the first of five new pressure suit prototypes had been delivered. NASA agreement on the shortcomings of the first Apollo pressure suit design was still three months away. Deciding what to do about the pressure suit would take even more months. While the life support side was successful in meeting all requirements, manned testing of the first new-design complete suit system had proven the Apollo requirements were still inadequate. This sent the life support side of the program into limbo until new, higher requirements could be agreed upon. While Hamilton and a few NASA engineers understood what the new backpack requirements needed to be, they had to convince NASA management in Houston that these were truly the correct requirements, and then the Houston management had to convince the management at NASA Headquarters. Formal direction for a backpack redesign would not come until 1964. The original Space Suit Assembly contract succeeded in identifying the challenges and requirements. Meeting those technical challenges were the next step.

CHAPTER 3

Recovery through Invention and Determination

At the time of the Apollo program the expression "back to the drawing board" was frequently used. A drawing board is a large, flat surface, usually a simple table where the drawing surface can be tilted to suit the preference of the engineer or draftsman. As the drawing board was the point at which most design efforts started, the expression meant starting over again from the beginning. This would have been a more desirable place than where the Apollo Space Suit Assembly program found itself in the spring of 1964 as the program proceeded under contract extensions.

The American nation expected NASA to deliver lunar explorations before the end of the decade. For that NASA needed an effective spacesuit to support spacecraft and mission development. Failure was not an option. Consequently, NASA took steps to assure the program's success. Hamilton Standard was NASA's prime contractor for the entire spacesuit system. From the beginning of its contract to NASA, Hamilton not only performed activities paid by NASA, but also took proactive steps under its own internal funding by recognizing risks to program success. As NASA had designated the International Latex Corporation, better known as ILC, as pressure suit supplier and supported ILC in its bargaining with Hamilton, Hamilton initially focused on life support issues and expected ILC to address pressure suit issues.

As Hamilton was the prime contractor to NASA, it was responsible for every aspect relating to the Apollo spacesuit. Regardless of the reasons, Hamilton had failed when it came to the pressure suit. This resulted in various NASA and Hamilton initiatives. During the original contract, ILC initiatives initially focused on a business-oriented agenda. In parallel, NASA and Hamilton made efforts to address the technical challenges facing the pressure suit. One of the interesting dynamics of the period was that Hamilton was under an obligation to share all related developments with NASA and ILC. However, NASA and ILC were under no obligation to reciprocate and usually elected not to share developments with Hamilton. This had programmatic consequences. In 1964 the effort to overcome spacesuit challenges grew to encompass almost all potential suit manufacturers. This resulted in increased competition for ILC and a realignment of that organization's efforts in the quest for man to be able to walk and work on the Moon.

EARLY PROACTIVE NASA, HAMILTON, AND ILC INITIATIVES

During the initial NASA-funded Hamilton Apollo spacesuit contract, NASA, Hamilton, and ILC conducted parallel activities to reduce the risk of being unable to meet technical or programmatic challenges. These efforts had impacts on developments as the program moved forward.

Starting with the Mercury program, heat rejection in space relied on liquid water exposed to very low pressure boiling at a reduced temperature. Boiling water very efficiently carries away thermal energy (a.k.a. "heat"). The problem with this approach was that it used wicks, chambers, and valves to control the boiling process. The wicks could freeze. The valves could collect condensation that would then become ice to jam the mechanism. Hamilton used this approach for their first Apollo backpack and successfully balanced the challenges to produce a backpack that met all requirements in initial testing. Hamilton management recognized the reliability risks that might not have yet been experienced. In parallel with the program activities of 1963, Hamilton internally funded research and development in two areas focused on more effective spacesuit heat removal. The first was how to better take heat from within the spacesuit and transfer it into space. The inventors who solved this problem were John S. Lovell and George C. Rannenberg III.

From the research and development that supported the Hamilton Apollo competition proposal of 1962, Lovell recognized that Hamilton's and the aerospace industry's approach of using water "boilers" with wicks and valves for transporting heat collected in the spacesuit and rejecting it to space had complexities that detracted from both reliability, efficiency, and compact packaging. His idea was to create a metal plate with microscopic holes that were just the right size so that water could freeze in the plate without damaging the plate. If this could be done, then the vacuum of space would automatically freeze the water in the plate sealing off the flow of water when heat removal was completed. If water below the plate collected any heat, then the water in the microscopic holes would thaw and the water would reach the space vacuum and boil or "sublimate," which would very effectively reject heat from the spacesuit to space. Once the heat was gone the holes in the plate would refreeze, automatically stopping the water and heat rejection. This would all happen with no moving parts or reliability issues. However, Lovell had a problem. While he was great at explaining what he needed, he was not good at developing hardware or making prototypes or test setups. Fortunately, John knew a scientist/engineer named George Rannenberg who was a genius at making things that worked.

In 1962 and 1963 Hamilton funded Lovell and Rannenberg's research.

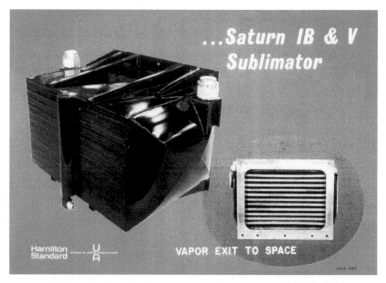

Figure 3.1. The first Apollo sublimator to reach flight
(courtesy UTC Aerospace Systems)

Together, they produced the Porous Plate Sublimator (U.S. Patent 3,170,303) which reached manned chamber testing in 1965. The sublimator would not only be used on the lunar spacesuits but also the rockets that carried Apollo crews into space on their lunar journeys, the Lunar Module (the Moon landing vehicle), and still see service in the U.S., Russian, and Chinese spacewalking spacesuits of today. The first sublimator to reach space flight (Figure 3.1) was for the Saturn 1B rocket.

The second area of research and development was the removal of body heat generated by the astronaut while working. At the beginning of 1963 the Hamilton concern was that astronauts would experience unacceptable levels of dehydration, as a result of sweating heavily, for adequate removal of body heat. To address this, Hamilton funded the work of David C. Jennings. Jennings' fitness, both physical and mental, was immediately apparent as was his passion for aviation and fascination with anything technical. When he spoke, there was no mistaking that he was a native New Englander. His calm tone and easy pace made it easy to follow his explanations. While Jennings' mild manner was his norm, he could be resolute as needed, like the combat infantry veteran of WWII he was. An aeronautical engineering graduate of Troy, New York, he was in all things professional.

Jennings' 1963 work started with air-cooling garments to more efficiently remove both heat and moisture from perspiration with minimum noise and gas flow around the head to irritate the eyes. However, this Hamilton effort was not the only cooling garment development.

In parallel with Hamilton's various efforts, NASA had recruited a distinguished Royal Air Force (RAF) flight surgeon named Dr. John Billingham to

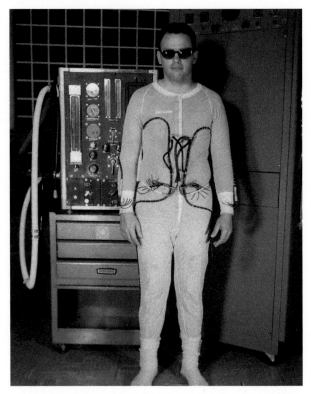

Figure 3.2. NASA's Jerry Goodman demonstrating the Crew System's Cooling Garment concept

(courtesy NASA)

head NASA's Environmental Physiology Branch in Houston. Billingham was aware that the RAF had started development of a full-torso cooling garment based on a water-cooling concept in their clothing laboratory at the Human Engineering facility in Farnborough, England. Billingham attempted to borrow an RAF example for an Apollo evaluation in the spring of 1963. Possibly as a result of it being classified by the British military, the arrival and evaluation were delayed until the fall. In the interim, Billingham and the folks at Crew Systems created their own prototype by June 1963 (Figure 3.2) based on what they knew of the RAF design.

The Houston garment was similar to the RAF design in having water-filled tubes and the torso was made of a solid fabric that effectively precluded air cooling. The RAF garment was made of a heavier fabric and the tubes were kept in place by weaving the tubes through small holes made in the torso fabric. The Houston unit had tubes sewn in place. The designs and evaluation results were not shared with Hamilton, either due to British security restrictions or out of a desire to see what Hamilton would develop in parallel.

In early October 1963 the development of Hamilton's cooling garment took a different path. Hamilton acquired the classified report on the liquid cooling vest effort conducted by the RAF in the early 1950s, which David Jennings was allowed to read. This led to Jennings' (Figure 3.3) experiment that resulted in

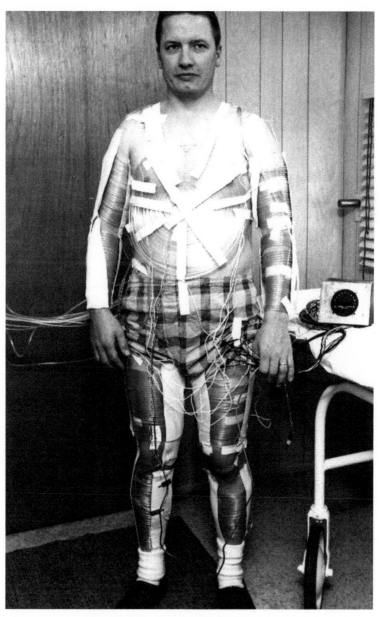

Figure 3.3. Harlan Brose ready for test
(courtesy UTC Aerospace Systems)

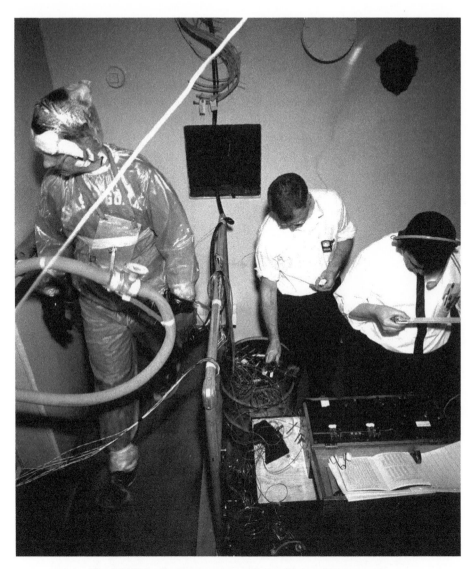

Figure 3.4. Concept testing with David Jennings (center)
(courtesy UTC Aerospace Systems)

the invention of the Apollo Liquid Cooling Garment (U.S. Patent 3,289,748). On October 15, 1963 Jennings wrapped 300 feet of 3/16-inch vinyl tubing around engineer and test subject Harlan Brose. Brose donned the cooling garment, was sealed in plastic to retain all perspiration (Figure 3.4), and covered with multiple layers of the warmest winter outer clothing available. The subject then exercised strenuously on a treadmill for periods up to two hours. For safety reasons, this and all subsequent cooling garment testing was conducted with a medical doctor present. More testing would be conducted through November 1963 to gain an understanding of body reactions to liquid cooling and liquid-cooling capacity.

Jennings' subsequent concept, like the RAF and Houston cooling garment designs, was based on circulating cooling water around the body. The torso garment held water-filled tubes against the suit user's body that allowed cooling water to collect and carry away metabolic heat. Like the Houston design, Jennings' concept had the tubes sewn to the inside of the torso assembly. The significant difference from the RAF and Houston concepts was that Jennings elected to create an open-mesh garment to additionally allow air circulation over the user's body for comfort, humidity removal, and additional cooling.

In parallel with manned concept testing, Hamilton was developing liquid cooling garment or "LCG" manufacturing processes, facilities, and staffing. While Hamilton was not a clothing manufacturer, its fiberglass aircraft component manufacturing processes utilized structural fabric techniques and employed seamstresses who cut and sewed fiberglass cloth layers to precise tolerances such as in pressure suit construction. The person selected to lead Hamilton's initial cooling garment manufacturing was a master seamstress named Rose Grady. Grady (Figure 3.5) would be responsible for the creation of all Apollo cooling garments in 1964 and for U.S. Air Force and commercial garments in subsequent years.

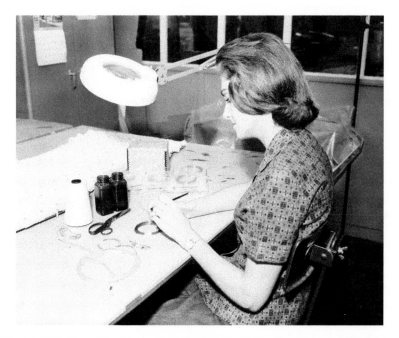

Figure 3.5. Rose Grady, the first Apollo Liquid Cooling Garment fabricator
(courtesy UTC Aerospace Systems)

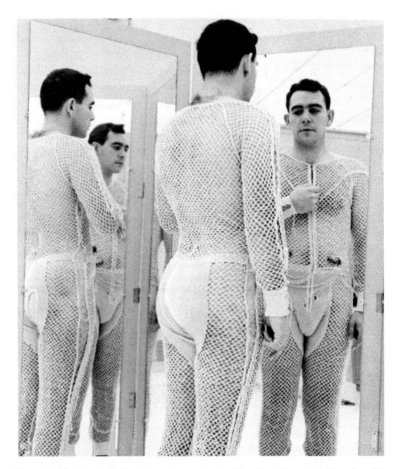

Figure 3.6. Mark Britanisky demonstrates Cooling Garment Number Two
(courtesy UTC Aerospace Systems)

The test results and subsequent analysis were used to create the first Hamilton liquid cooling garment. The completion of assembly and first test was in December 1963. The first prototype was a two-piece fishnet-like, cotton mesh garment to which 232 feet of tubing were carefully attached via hand sewing. During the spring of 1964, this cooling garment was extensively tested in Windsor Locks, Connecticut and in Houston. It proved to be highly successful. The analysis of the performance provided the basis for the revisions that flowed into the next cooling garment prototype. Hamilton's second (Figure 3.6) and subsequent cooling garments were one-piece full-torso garments, as would be the cooling garments used on the Moon. For the second prototype, the tubing was increased to 267 feet and the tubing pattern revised for more efficient heat removal. This second design began and completed manufacture in April 1964. This supported manned testing in Windsor Locks and Houston in the coming months.

Hamilton's first exposure to outside cooling garment developments came in June 1964 with a visit by NASA's Gil Freedman and Frank DeVos to Windsor Locks. Hamilton was given the opportunity to review the latest RAF prototype in relation to its simple technique of weaving the tubes through holes in the torso garment to hold the tubes in place. While it offered potential cost savings to the Apollo program, as it was less labor intensive than Hamilton's sewn-on method at the time, Hamilton pointed to evidence that the tubes on the RAF garment had slid in the garment and kinked during previous use, thus shutting off cooling flow in those loops and reducing overall cooling ability. The RAF technique was not subsequently adapted into the Apollo program.

Hamilton's third cooling garment prototype made its debut on June 25, 1964. The amount of tubing and cooling loops in this prototype was increased. This third garment remained at Hamilton as the program's principal test item. Testing typically now involved Arctic-type clothing (Figure 3.7) to provide spacesuit-level heat retention.

Hamilton's favorite cooling garment test subject was Frank Piasecki (Figure 3.7) as a result of his ability to perform five-hour treadmill test sessions. During the summer and fall of 1964, Rose Grady created two more design iterations and at least three more prototypes for the Apollo program. These efforts were intended to produce a commercial line of products as well. July 1964 brought the debut of Hamilton's NASCAR Driver Cooling System. Paul Goldsmith wore a NASCAR cooling garment and stayed comfortable and alert even though track-level temperatures reached 130 degrees during the race.

In parallel with backpack and cooling garment developments, Hamilton internally funded further development of internal pressure suit development capabilities as well. In April 1964, Hamilton hired Dr. Edwin G. "Ed" or "Doc" Vail. Vail was a B.F. Goodrich Air Force pressure suit technical consultant who had also supported NASA on the Mercury program. To reflect the importance Hamilton placed on this recruitment, Vail was provided an office in the top executive area that was commonly called "Mahogany Row" because of the mahogany wood trim that adorned the offices, conference rooms, and adjacent areas. Vail immediately pushed the desk into the hall to make room for sewing machines and a cutting table for training suit assembly personnel. This probably did not endear him to the adjacent top Hamilton management. Lab space was soon found and Vail and his operations were relocated.

Vail inherited the Mark Baker group that had produced the Hamilton "playsuit" and was given the opportunity to recruit anyone he wished internally and externally. He immediately recruited an experienced Air Force suit technician and toolmaker named James "Jim" Hopper. Vail had previously worked with Hopper. Internally, he selected Andy Hoffman as project engineer, Ed Brisson as primary suit subject, and Harvey Smith as mechanical designer.

Vail soon arranged to borrow a David Clark X-20 Dynasoar program suit from the Air Force for evaluation at Hamilton. X-20 was a short-lived program to develop and deploy reusable space vehicles launched atop a rocket from a launch pad but featuring more conventional landing gear to allow Space

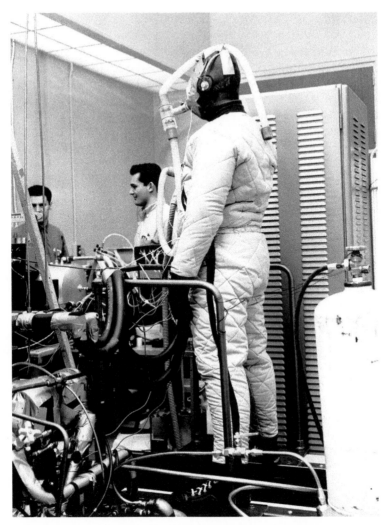

Figure 3.7. Frank Piasecki testing Cooling Garment Number Three
(courtesy UTC Aerospace Systems)

Shuttle-style runway returns. The X-20 suits used the latest in David Clark mobility technology and had an excellent reputation for reliability. As there were intellectual property issues, Hamilton prevailed upon NASA to gain David Clark Company cooperation and support for the evaluations.

With Jim Hopper coming on board, Vail thought that he and Hopper could find a solution to Apollo suit mobility deficiencies. As Vail and Hopper preferred to work separately, this effectively created a second Hamilton suit development group under Baker. Those who had not effectively contributed during Baker's earlier playsuit effort were replaced by the incremental hiring of young, new engineers such as Michael A. "Mike" Marroni Jr., Doulas E.

"Doug" Getchell, and John C. Hardy. Given the lack of success with Hamilton's first pressure suit developments, Mark Baker took a step back with his small group and started methodical evaluation of brainstorming mobility design approaches, making elbow, knee, or brief prototypes and then testing them on a device that measured bend effort that could be compared to existing ILC Apollo designs. This approach did not provide immediate success.

As pressure suit mobility had proven such a challenge, NASA additionally explored Litton's capabilities and technologies starting in 1963. The young NASA engineer in charge of this effort was Joe Kosmo. In the late 1950s Dr. Sigfried Hansen Litton had developed a pressure suit to speed vacuum tube development (Figure 1.17). Up until now, the U.S. approach to pressure suits had usually been fabric structures with zippers for entry. NASA's Manned Spacecraft Center in Houston funded Litton to build on the Mark I technology and construct a prototype to determine the applicability of this approach for Apollo. The resulting prototype was the RX-1 (Figure 3.8).

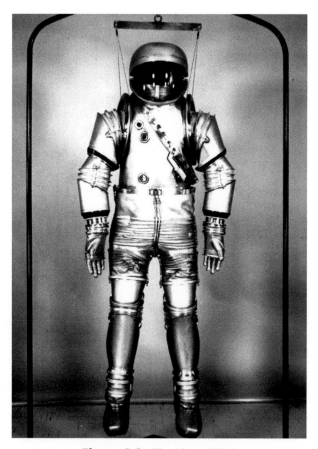

Figure 3.8. The Litton RX-1
(courtesy UTC Aerospace Systems)

The "RX" designation was not assigned by Charlie Lutz, as were all the other NASA Apollo suit designations, but the product of a random office discussion. "RX" appears on the header of American doctor prescription pads, which is an abbreviation of the Latin word for "recipe." As Joe Kosmo was the NASA engineer in charge of this development, he felt responsible for naming the effort. Kosmo happened to be joking with his office mate, Jack Rayfield, about the first Litton suit being the prescription that would solve all the frustrating problems being experienced on the Hamilton/ILC Apollo suit program. In the inspiration of the moment, Kosmo turned and wrote RX on a chalkboard on the wall. Kosmo liked what he saw and presented the designation to his management who accepted his recommendation.

Dr. Hansen had left Litton in 1959, creating the opportunity for someone new to lead the creative effort of the RX-1. This was initially G. Fonda Bonardi. Later, the responsibility for Litton RX-1 development passed to William Elkins, who joined the Litton Space Lab in 1963. The resulting first Litton pressure suit prototype weighed 83 pounds (37 kg) and featured revised mobility systems with significant improvement in mobility.

During this period, ILC internally funded efforts to reduce Apollo pressure suit production costs. Moreover, ILC requested Hamilton funding to explore development of a David Clark Gemini-style rear-entry system. As ILC's management had chosen to ignore Hamilton-identified concerns and these activities were not issues to Hamilton, they further complicated functioning of the Apollo program.

However, these early efforts were just the beginning. They did not yet supply the needed solutions. Thus, the community efforts continued and expanded until success was achieved. The path forward had many unexpected twists and turns.

CONTINUED STRUGGLES AND NASA DIVIDING THE APOLLO SPACESUIT INTO THREE "BLOCKS"

March 1964 saw the darkest days in the Apollo spacesuit program. Command Module testing found that the David Clark Gemini suit was the only one acceptable. Lunar Module testing was worse as the suit evaluators unanimously rejected the latest ILC Apollo design with Astronaut Gordon Cooper emphatically elevating his dissatisfaction to the attending Washington and Houston NASA management. Almost everyone understood the magnitude of the problems and that there would be significant consequences.

Immediately following Grumman's Lunar Module evaluations, the AX3H suit was sent back to Hamilton. Hamilton conducted an evaluation with the greatest of haste. In parallel, Hamilton's President Bill Diefenderfer attempted to contact Wally Heinze, the President of ILC. For over a year and a half, the means of resolving all issues between Hamilton and ILC were communications between Wally Heinze and Bill Diefenderfer. Heinze was on vacation with his family in Europe and left strict orders that he was not to be disturbed. As far as Diefenderfer was concerned, there was a pressure garment mobility and reliability crisis. No matter how vehement Diefenderfer became, ILC's Vice President and Treasurer Irving Obrow would not attempt to reach Heinze.

Testing at Hamilton indicated that the latest ILC suit had less mobility than the previous design iteration. The AX3H suit was then shipped to ILC for quality inspection and remanufacture. In parallel, Hamilton sent a five-person task force to ILC that was headed by a quality assurance engineer named Robert E. "Bob" Breeding. At five foot ten inches with an athletic build and a broad, flattened nose, one could easily get the impression that at some point he had been a boxer. His demeanor could be whatever was required. He could be soft spoken or he could bark orders, he could be most amiable and an excellent host or he could be intimidating and overwhelming. Breeding's career took a longer path than most in that he served eight years in the Army before joining Hamilton in 1953. He was not afraid to assert himself in a situation. Married with children, Breeding took a Fire and Security Department job working mostly evenings, nights, and weekends so he could further his education part-time during the day. By 1962 Breeding had earned his Bachelor's Degree from the University of Hartford enabling him to gain a position in quality assurance at the outset of the Apollo spacesuit contract. In less than two years, Breeding demonstrated an excellent understanding of what NASA expected and an ability to lead. In critical situations, his "take-no-prisoners" approach did not especially make him liked, but he certainly gained the respect of all.

Deployment of the five-person Hamilton task force marked a change in the Hamilton/ILC relationship. In the first 14 months of working together Hamilton had accepted a "hands-off" relationship. In the beginning ILC had claimed they were capable of meeting all contractual requirements. Hamilton had no reason to initially question that claim. Prior to the task force, Hamilton personnel coming onsite at ILC's Dover Delaware facility had to be prearranged through Hamilton's Purchasing Department and ILC's Apollo Contracts Representative. Hamilton had to provide a reason for the visit and ILC had to concur that the reason was valid. With the task force in place, Hamilton's management issued notifications directly to ILC management with "no" not being an acceptable answer. Fortunately for everyone, the ILC-Dover management, unlike their New York superiors, understood the urgency. ILC-Dover accepted and immediately started working with the Hamilton group.

In parallel with deployment of the task force, Diefenderfer sent formal correspondence to Heinze. Heinze's first reply appears to show that he still did not fully understand the state of the program as he still expressed the opinion that the pressure garment issues "were being greatly exaggerated."

The mission of the task force was to revise the third suit design into a sufficiently acceptable configuration to allow new training suit production which NASA desperately needed. The Hamilton personnel selected for the task force were those with the greatest knowledge of pressure garments. Moreover, such experience fell short of that of the ILC personnel with whom they interfaced. However, it was NASA's perception that the preceding problems were the result of Hamilton's failure to manage ILC and NASA expected an immediate quality review and improvement of the design.

As far as the ILC technical personnel in Dover were concerned, the Hamilton task force was probably an impediment to progress in that it laboriously analyzed every facet of the AX3H suit to document all suit issues and identify root causes before progressing to corrective actions. This reflected a cultural difference. The ILC people thought this was painfully slow. They could have revised and tested the suit many times over in the time Hamilton accomplished one review and remanufacture. As far as the Hamilton group was concerned, the slow methodical process represented the good engineering practices expected by NASA management, who were displeased. Thus, the ILC-Dover personnel had undesired "help." However, this was not the only problem these events posed for ILC. In April 1964 NASA began openly discussing the use of Gemini suits for the first Apollo manned missions, which did not have "spacewalks."

By late May 1964 the latest ILC Apollo prototype had been remanufactured with Gemini-style life support connectors and a Hamilton-designed and manufactured "Universal" helmet (Figure 3.9). Except for a glove failure, all testing in the next month was favorable. This allowed ILC to resume Apollo program development. In June 1964 NASA directed the resumption of training suit manufacturing based on the revised features. To differentiate this training suit production from the negative evaluations of the original AX3H suit, NASA later directed the new training suits be designated AX4H until the first delivery was accepted and then A-4H to indicate fourth-generation "production" suits.

Upon return to Hamilton from Dover, the Breeding-led task force received great praise from ILC's President Wally Heinze. This praise, which was probably limited to the New York City headquarters, was visible to NASA. Hamilton's activity also drew positive feedback from Dick Johnston and others within NASA.

Accompanying the introduction of the Universal Helmet was the Apollo program's second and better known Emergency Oxygen System or "EOS." This had the same operating pressure and five-minute performance but was lighter and much more compact. This EOS was a bagel-sized and shaped device with a simple regulator and activation systems. Packaging studies of where to install the EOS had resulted in it being mounted on the rear of the helmet. This rear-mounted EOS concept was probably the more interesting in that it permitted utilization of the unused space behind the helmet and above the backpack.

Immediately after completion of the remanufactured AX3H suit, ILC delivered a separately donned thermal micrometeoroid overgarment as well as test samples of the overgarment layers and construction details to support parallel testing. The samples met the thermal requirements but not the impingement requirements. The premise driving impingement requirements was that an astronaut should be able to survive ejecta thrown into space by a

Figure 3.9. AX3H-024 suit as delivered to NASA June 1964
(courtesy UTC Aerospace Systems)

meteorite strike and then drawn back to the surface of the Moon by lunar gravity.

The remanufactured AX3H went on to be used with the overgarments in system-level field evaluations conducted in Bend, Oregon (Figure 3.10). This location was selected for its similarity to the lunar landscape. The evaluations included Astronaut Walter Cunningham, who was part of the Apollo 7 first manned mission. Movement over uneven terrain in the pressurized suit was difficult, and the thermal overgarments were found to be an encumbrance and had layer separation issues. This only reaffirmed NASA's view that the Apollo suit design was not yet adequate for lunar exploration.

The delivery of the remanufactured AX3H suit was also the milestone for ILC to start A-4H Training Suit production and resume pressure garment development. In parallel, Hamilton internally funded suit development activities. This was principally focused on learning about the dynamics of pressure garments.

Figure 3.10. AX3H field testing conducted In Bend, Oregon
(courtesy NASA).

In the opening days of June 1964, Wally Heinze tried to elevate the disagree-ment between ILC and Hamilton by attempting to contact Bill Diefenderfer's boss, William P. Quinn, the President of the United Aircraft Corporation. On June 6, 1964 Diefenderfer was required to give a presentation to Quinn on the contract performance of ILC in East Hartford, Connecticut. Upon Diefen-derfer's return to the Windsor Locks facility, he immediately ordered establishment of a dialog with B.F. Goodrich for it to provide additional Apollo pressure suit development and technical support under Hamilton inter-nal funding. Quinn elected to initially ignore Heinze. A week later, Heinze and Irving Obrow were able to "run into" Quinn while he was visiting NASA in Houston. However, Heinze was unable to cultivate a favorable relationship with Quinn. In parallel, Goodrich signed a support contract with Hamilton. The first Goodrich prototype arrived in Windsor Locks in September.

Another outcome of the meeting between Heinze and Quinn was the announcement by ILC that they were making an organizational change aimed at strengthening their efforts on pressure suit development specifically; to this end they hired Dr. Nisson Ascher "Art" Finklestein. While Finklestein knew little about spacesuits, he brought management and development program experience plus a demeanor that Hamilton found more accommodating. This perhaps stemmed from his family life. Prior to joining ILC, Finklestein had helped his wife Rona take care of their house and two sons while he worked and she earned her Doctorate degree at the University of Rochester, which she finished just days before Finklestein joined ILC.

At about the same time (June 1964), NASA had become concerned at the pace of Hamilton backpack development. NASA felt that if Hamilton could produce a successful backpack in 10 months, then they should have been able to "tweak" the system for greater capacity in four or five months. They did not accept Hamilton's view that the increased capacity with the same volume and weight meant they had to go "back to the drawing board." The Hamilton estimate of early 1965 for a man-rated prototype was not well received and triggered NASA to fund the AiResearch Division of Garrett Corporation located in Torrance, California, which is now a unit of Honeywell, for continued development of its Personal Environmental Control System. The Personal Environmental Control System or "PECS" was initially developed under the Gemini program but had lost funding. This permitted NASA to have a second backpack system being developed in parallel, in case Hamilton failed in its second backpack development.

To facilitate final inspection acceptance of training suit production and oversee Apollo development, NASA had deployed engineer Robert L. "Bob" Grafe to Windsor Locks, Connecticut. Grafe provided weekly memos to his boss Matt Radnofsky. The memos provided the status of all Apollo pressure suit developments being performed at Windsor Locks, Dover, and Akron.

At the start of September 1964, Hamilton replaced the mild-mannered Roger Weatherbee with Bob Breeding as the Hamilton Apollo Program Manager. Encouraged by positive feedback from ILC's top management and NASA's high regard for the Breeding-led Hamilton AX3H task force, Hamilton management probably hoped that Breeding's more assertive and versatile style might avert the program changes that NASA was considering.

By September 1964 the Hamilton-funded Space Lab in Windsor, Locks had been completed. This included a 10×10-foot Man Test Chamber designed specifically for manned testing under space vacuum conditions. It contained a treadmill plus various biomedical and suit test equipment. The chamber was capable of full-altitude (space) vacuum and temperatures of $+300$ to -65°F.

To assure test subject safety a medical treatment facility that included a portable hyperbaric chamber was put in place. The lab additionally featured smaller thermal and vacuum test rigs that could support above 300°F down to -200°F test cells that could simulate launch vibration and shock loads. Moreover, there were test rigs designed to fully verify Apollo Space Suit Assembly and Lunar Module life support requirements at the component level. To assure test subject safety and gain maximum knowledge from manned testing, Hamilton recruited Dr. Vance H. Marchbanks Jr. Marchbanks was the son of an African American career soldier who enlisted as an Army cavalryman and rose to the rank of captain in World War I. Dr. Marchbanks earned his bachelor's degree at the University of Arizona and earned his medical degree from Howard University. In 1941 he was a staff member at the Veterans Hospital in Tuskegee, Alabama when patriotic duty called him to join the Army Air Corp. Marchbanks served as flight surgeon in Italy to the 332nd Fighter Group, better known as the "Tuskegee Airmen," where he was awarded the Bronze Star. After WWII, Marchbanks elected to stay in what was then the Air Force where he received further distinctions and decorations before being selected as one of the original 11 NASA medical doctors in 1960. This NASA service included Marchbanks monitoring John Glenn's biomedical readings to assure his wellbeing during the first U.S. orbital space flight.

In 1964 Marchbanks left NASA and the Air Force (as a colonel) to join Hamilton (Figure 3.11). At Hamilton he gained the nickname "the Mercury Medic." Marchbanks oversaw manned testing and provided input to Apollo-detailed specifications, thus aiding the operational success of Apollo. Beyond his Apollo program duties, he also ran Hamilton's Medical Department on the

Figure 3.11. Vance Marchbanks and Hamilton's almost completed Man-Vacuum Chamber

(courtesy UTC Aerospace Systems)

campus at Windsor Locks, Connecticut overseeing the health of Hamilton employees while they were at work. He was universally respected and liked.

By the end of September, it was clear that NASA was reorganizing the Apollo Space Suit Program. Details of the program changes were formulated in NASA between Washington and Houston. The beginning of October 1964 had special meaning for NASA as it marked the second anniversary of the start of the Apollo Space Suit Program; whereas in early 1962 development had been expected to take 10 months. NASA recognized the program would reach its second anniversary without an acceptable Apollo spacesuit being successfully developed. Pressure suit mobility development had yet to meet the challenge and manned testing of a second Apollo "backpack" design was nowhere in sight. The Apollo program could not afford the limited progress of the first two years to continue. The NASA schedule put immense pressure on the Crew Systems Division and its contractors. To assure schedule compliance and ultimate success, NASA split the Apollo spacesuit effort into three parallel paths. NASA announced that the Apollo pressure suits would be divided along vehicle system linesthese were "Blocks I, II, and III."

Block I systems allowed reaching Earth orbit and proving out the Command Module. The David Clark Company was to be contracted for Gemini-based suits for the early Apollo Block I orbital missions that did not include Lunar Modules and spacewalks.

Block II was to prove out the vehicle system that included a configuration of the Lunar Module to permit walking on the Moon and accomplish America's goal of setting foot on the Moon before the end of the decade. To achieve this, Hamilton's Apollo contract would be continued but with the early and late missions removed. NASA saw the forthcoming A-4H training suit production as a temporary improvement. NASA expected subsequent designs to meet all Apollo requirements. The Block II spacesuit program would have two significant milestones in its first year. By December 1964, Hamilton had to provide a pressure suit design recommendation that met all requirements to support production of a subsequent "A5H" training fleet. As NASA did not have confidence that the A5H configuration would be adequate for lunar operations, it would conduct a Block II pressure suit competition in June 1965 to establish the pressure suit configuration that would be used on the Moon. To support NASA's timeline, Hamilton scheduled comparative testing of all possible Apollo pressure designs in November 1964.

To distance itself from the program's first two years, NASA implemented a linguistic change. In the late summer, there was a semi-formal name competition within NASA. The suit system name was revised from the Space Suit Assembly to the Extravehicular Mobility Unit or "EMU." The Pressure Garment Assembly became the Pressure Suit Assembly or "PSA."

Under the Block II reorganization, NASA's James "Jim" O'Kane and Dr. Robert L. "Bob" Jones gained funding for internal NASA development of a full polycarbonate bubble that attached directly to the neck ring. This would eliminate the intermediate fiberglass shell and associated bond joints. The two

NASA operatives were a study in contrasts. O'Kane was a serious, meticulous, detail-driven engineer, while Jones was a quick-witted, joking psychologist. The mix seemed to work, much to the amazement of their peers.

The Block III spacecraft planned for the late Apollo missions had additional capacity to permit riding on or flying over the Moon for exploration that was not limited to walking distances using limited life support durations. It was recognized that the Block II pressure suit being used for launch, reentry, rescue, and extravehicular activity or "EVA" resulted in functional compromises in each use. Block III was to develop a dedicated, better performing lunar surface exploration suit system. NASA Crew Systems were tasked with performing system-level development and managing contractor suit and backpack contract performances.

In October 1964 a Hamilton delegation that included President Bill Diefenderfer and Apollo Block II Program Manager Bob Breeding met with Dick Johnston, Chief of NASA's Crew Systems Division, to understand NASA's expectations for what had become the early extravehicular portion of the Apollo program. NASA's Matt Radnofsky was also present. Hamilton accepted the expected developments and timeline that would lead to the selection of an A5H design but expressed concern regarding the Block II pressure suit competition. Johnston assured Hamilton that as long as the Hamilton entrant met all the Apollo requirements, Hamilton would remain the Block II prime contractor regardless of the pressure suit winner. However, no one in the meeting likely envisioned the events that would follow.

HAMILTON'S BLOCK II SPACESUIT PROGRAM

As far as NASA management was concerned, Hamilton's initial mission was to produce a fleet of A-4H training suits and completely develop the pressure suit and backpack to be used on the Moon. NASA expected this to be accomplished in less than one year. The simplest part of this challenge should have been A-4H training suit production.

NASA directed it would select the winning design and manufacturer for the subsequent pressure suit design that would be used in the early Apollo lunar missions.

ILC's 1964 Training Suit Production

By the end of August 1964 the first ILC A-4H torso assemblies had been readied at Hamilton to be married with Hamilton helmets and tested as pressure suits for delivery to NASA (Figure 3.12). To facilitate deliveries, NASA elected to have its own personnel stationed in Windsor Locks, Connecticut to perform acceptance inspection at Hamilton. These first A-4H suits proved to be an embarrassment to both Hamilton and ILC. The torso assemblies were plagued with minor problems causing the initial units to be returned to Dover

Figure 3.12. An early A-4H suit with the Universal Helmet
(courtesy UTC Aerospace Systems)

as quickly as they were being delivered. As the backlog grew, Hamilton began fixing the A-4H suits in their new Windsor Locks Suit Lab.

Hamilton-made helmets were also having problems. Hamilton helmets, like their ILC predecessors, had "drop-down" pressure visors. When the suits were unpressurized the visors were flipped up and protected inside the helmet's shell. To pressurize the suit the visor flipped down and moved forward to contact the pressure seal inside the face area of the helmet. The visors in the Hamilton helmets unpredictably worked and stopped working. This could be more than embarrassing. The astronaut would usually have the pressure visor up so he could easily breathe cabin oxygen. If an emergency arose where the astronaut needed to pressurize his suit and the visor did not drop and seal, he could die. This visor problem was very confusing to Hamilton as their first two prototypes delivered to NASA were working flawlessly. The reason for the first two successes proved to be the helmets being manufactured and assembled by skilled toolmakers. The subsequent units were manufactured and assembled by production personnel. While the helmet problems were corrected relatively quickly, it was not fast enough to avert formation of a movement within NASA for A-4H suits to adopt a new Hamilton-designed helmet being planned for the A5H configuration. This was a "fixed"-type helmet, meaning that the pressure visor was permanently bonded to the helmet shell.

At this juncture, NASA desperately need latest-possible configuration training suits and intended the A-4H configuration to be a limited production item without changes as a stopgap effort because the suits did not need to meet Apollo requirements. NASA had mandated the subsequent "A5H" design of training suits would meet all requirements and scheduled the start of their delivery for March 1965. There were contracts and presentations identifying the A5H suit as a new, completely compliant design. Consequently to accommodate the NASA requested helmet change to the A-4H suits that were already in production, the first A-4H suit with the "old" helmet became "early" A-4H suits. The design change with the new, fixed-type helmet was called "late" A-4H. Five early A-4H suits were made. At NASA's behest some or all of these could have been made with a white training-type outer fabric like the AX3H suit. However, NASA ordered all these early A-4H suits with an aluminized flight-type outer layer (Figure 3.12).

After months of problems with A-4H torso assemblies the first A-4H suit was allowed to be delivered on December 11, 1964 at the personal insistence of Matt Radnofsky who felt NASA's inspectors were nitpicking. The suit was sent to Grumman and then immediately returned to the Apollo Space Suit Program for quality issues. The next A-4H delivery appears to have been under provisional acceptance by NASA on January 1, 1965. The other early A-4H suits soon followed.

At the beginning of 1965 NASA began phasing out the Xs and hyphens in program suit model designations so the A-4H became A4H. The next NASA-planned suit design would be the A5H.

In parallel with mobility development, there were events that drove helmet changes. At the beginning of 1964, all U.S. space helmets had pressure visors made of acrylic plastic; moreover, the helmet would fit tightly on the astronaut's head. Most helmets incorporated a pressure-sealing neck bearing to allow the head to rotate with head movement and a neck mobility system to allow looking down and up when pressurized. In 1964 a Gemini training accident had caused an acrylic helmet pressure visor to fracture. This resulted in NASA mandating polycarbonate pressure visors. Polycarbonate had far superior impact properties but no one had developed the ability to form polycarbonate to Apollo requirements for optical quality. Another desire was to remove head-borne loads. In normal 1-G and multi-G situations, G-forces are applied to the user's neck. If not adjusted correctly the suit pressure load tries to pull the tight-fitting helmet off the wearer's head. Movement could result in changes in suit volume/pressure applying force to the neck and head causing discomfort. Yet another desire was to eliminate the weight of the pressure visor retraction mechanism and the potential for visor malfunction. Integration of the

Figure 3.13. Apollo training spacesuit in a mission simulation in 1964
(courtesy UTC Aerospace Systems)

pressure visor into the helmet shell alleviated those issues. Apollo was the first space program to certify and implement a helmet with a polycarbonate pressure visor that exhibited satisfactory optical propertiesthis was the Hamilton C3 helmet.

The C3 helmet (Figure 3.13) met NASA's requirements of the time. This was a roomy fixed-type helmet that had no neck joint or bearing. The system impacts occasioned by the helmet necessitated the introduction of the Hamilton "bump cap" Communications Carrier Assembly. Unlike preceding helmets that had the communication microphones and earphones integrated into the helmet shell, this design required the communications equipment to be placed onto the head of the wearer. The Hamilton nickname of bump cap stemmed from it being padded. This nickname did not last long. As the Communications Carrier Assembly was similar in appearance to the headgear of a popular cartoon character, this soon became more commonly known as the "Snoopy" cap after the lovable cartoon character who daydreams of being a World War I aviator. This nickname has endured through the decades. Another feature of the C3 was that the Emergency Oxygen Supply or "EOS" was optionally attachable to and removable from the back of the helmet.

The C3 helmet had a larger neck ring. To accommodate this change, the upper torso of the A4H suits, which placed a greater burden on the program. The A4H production consisted of five early-configuration and seven late-configuration suits. By the summer of 1965, quality and delivery issues had

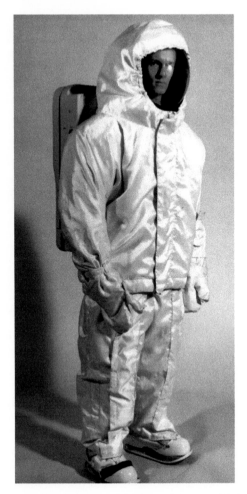

Figure 3.14. ILC Apollo A4H Thermal Meteoroid Garments
(courtesy UTC Aerospace Systems)

caused NASA's Bob Grafe to be placed at Hamilton continued to oversee the reworking the suits and monitor progress daily.

The last deliveries associated with A4H training suits were the thermal overgarment prototypes (Figure 3.14) delivered in June 1965. The garments were even heavier and more restrictive than the previous ones (Figure 3.10). This caused NASA to rethink their entire Apollo overgarment needs.

Developing the Backpack Used on the Moon

Developing the life support system for the Apollo 9 through 17 manned missions involved more than designing and making a backpack to allow people to explore the lunar surface. There was training on Earth that required astronauts while sealed up in pressurized suits learning how to remove carbon dioxide, add oxygen, control humidity, and provide cooling. The encumbrance of a life support umbilical not only detracted from the realism of NASA simulations, but also in many cases, such as Lunar Rover simulations, at times precluded effective evaluations entirely. In response, NASA elected to have Hamilton produce a simple compressed air and ice cooling backpack with an operating duration of 30 minutes to support vehicle simulations requiring spacesuits (Figure 3.13).

In the trade studies that led to the 1962 contract win, Hamilton had conducted research into the possibility of using a very simple cryogenic oxygen system for the lunar backpack. Oxygen is normally a gas that makes up 21% of the air we breathe. Cryogenic oxygen is superchilled until it becomes a liquid. When the oxygen is allowed to flow into the ventilation system of the spacesuit, it becomes a gas that cools the surrounding ventilation gases, thus automatically providing cooling. By feeding the oxygen into the ventilation system through an "ejector drive" device, the oxygen flow drives ventilation gases away from the ejector, thus pushing the gas flow through the suit without the need of a fan or a battery. When greater life support capacity in essentially the same volume became an Apollo program priority in 1963, Hamilton internally funded further cryogenic research and development.

Anticipating that having to stop to recharge a "training backpack" every 30 minutes would be a significant hindrance to NASA evaluations, Hamilton elected to fund the development and delivery of a longer duration, fully autonomous ground training backpack (Figures 3.15 and 3.16). This system provided cooling, humidity control, pressurization, ventilation, carbon dioxide removal, communication, and suit pressure for up to one and a half hours to simulate suited spacewalking conditions without the risk or expense of a staffed vacuum chamber.

Called the "Cryo-Pack" this not only permitted Earth-based pressurized training without umbilicals, but also offered practice operating the flight backpack controls expected to be used for Apollo. Prior to 1967 there was no provision for a chest-mounted Remote Control Unit as was used in the lunar missions. In 1964 the controls were located on the bottom-right corner of the backpack (Figure 3.17), which were out of the astronaut's line of vision. Control adjustments were performed by feel and experience.

In the fall of 1964 Hamilton made the decision to subcontract Apollo cooling garment manufacture. The selected vendor was B. Welson & Company of Hartford, Connecticut. Beyond the expected goals of this action, B. Welson brought new talent and softgoods-manufacturing experience to the effort. New materials and processes were introduced. A foreman at B. Welson named Stan Krupinski had the idea of using a buttonhole sewing machine to sew the tube attachments to the garment. This reduced the labor content and thus future production costs substantially.

The B. Welson Lead Seamstress, Giulia Yacone, would add many ideas and refinements. The first B. Welson cooling garment was produced in December

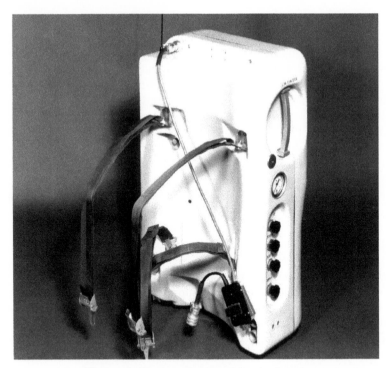

Figure 3.15. Cryo-Pack training backpack
(courtesy UTC Aerospace Systems)

1964. This was the sixth Hamilton design process, but the first Apollo cooling garment not only to feature machine-sewn tube attachments to the garment, but also the use of an elastomeric mesh, another B. Welson innovation. The elastomeric mesh helped keep the garment tight fitting yet comfortable and thus became the standard for the remainder of the Apollo program.

In January 1965 testing of the first B. Welson cooling garment resulted in refinements and the creation of yet another cooling garment prototype. To show NASA that the garment provided adequate comfort, Jennings wore the second prototype under his normal clothing for 14 continuous days in February 1965. The test was timed in such a way that Jennings could be present at a Liquid Cooling Garment review in Houston on the 14th day. When a NASA representative questioned the credibility of wearing such a garment for an entire 14-day mission, Jennings jumped up and pulled off his jacket, shirt, and tie, in his best impression of Clark Kent turning into Superman, to announce that he had worn the now-visible cooling garment for 14 days. To document that wearing the garment continuously for two weeks had no ill effects, Dr. Marchbanks carried out a physical examination on Jennings (Figure 3.18).

However, this seventh prototype was not the end of developments to maximize comfort. Eleanor Jennings, Dave's wife, made her own contribution to the Apollo program by proposing a chiffon comfort liner for abrasion pro-

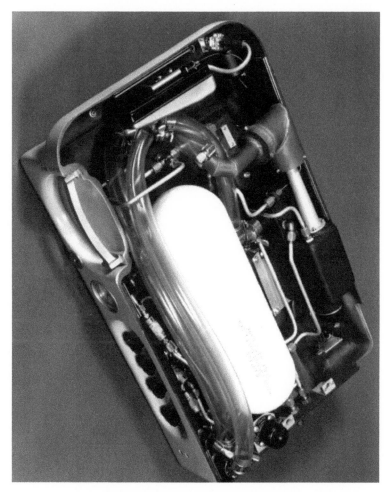

Figure 3.16. Cryo-Pack inside view
(courtesy UTC Aerospace Systems)

tection. She was neither an engineer nor a paid member of the space program. Her formal education ended with Nursing School, but instead of becoming a nurse she married Dave. Together, they had five children. Raising their children was Eleanor's full-time job.

The liner concept was incorporated in the next prototype (Figure 3.19). Subsequent testing indicated that heat transfer impairment as a result of the liner was negligible. Thus, all subsequent Apollo cooling garments used chiffon liners, which continues in U.S., Russian, and Chinese cooling garments right up to the present. While earlier B. Welson-made cooling garments were in shades of blue as a result of material availability, the eleventh was made using a natural (white) elastomeric material, thus establishing the tradition of white cooling garments that would be used on the Moon.

Figure 3.17. Apollo backpack controls in 1965
(courtesy UTC Aerospace Systems)

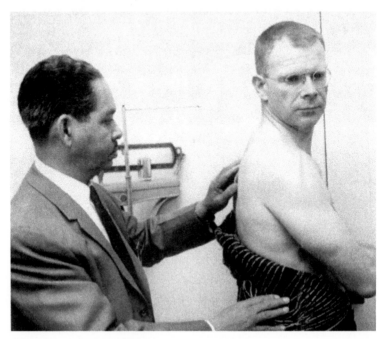

Figure 3.18. 3.18 Vance Marchbanks examines David Jennings after 14-day test
(courtesy UTC Aerospace Systems)

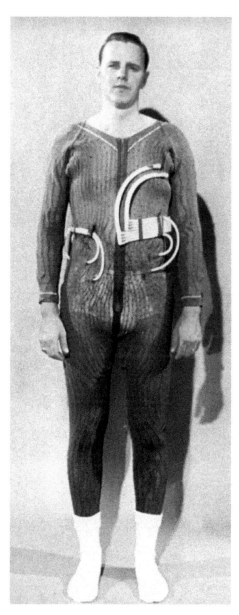

Figure 3.19. NASA's Frank de Vos demonstrating the eighth-design Apollo Cooling Garment

(courtesy UTC Aerospace Systems)

Proceeding in parallel with Hamilton's Block II life support system, NASA funded Block III lunar backpack efforts by AiResearch. This became a race to see which organization would complete successful manned testing and certification first. Hamilton was in the lead but any failure could change that.

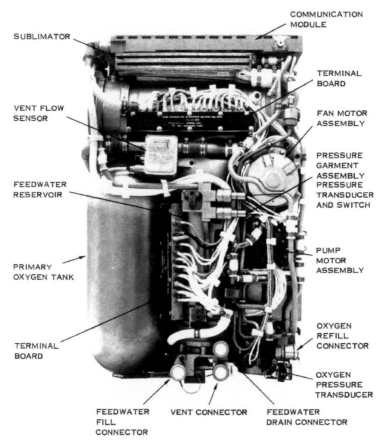

Figure 3.20. "Liquid-cooled" PLSS in 1965
(courtesy UTC Aerospace Systems)

The challenge involved in creating the Hamilton backpack (Figure 3.20) was not limited to the invention of liquid cooling garments and porous plate sublimators. Higher capacities and more components were needed to provide a crewmember with a liquid cooling system, both of which added weight and volume. Additional volume allowances were not available in the Apollo program. Thus, maximum performance had to be obtained from every component and subsystem.

By September 1965 Hamilton had successfully completed the unmanned testing and documentation needed to safety-certify the new backpack for manned vacuum chamber testing. This testing was conducted under the watchful eyes of Hamilton's Dr. Marchbanks and NASA Astronaut Dr. Joseph P. Kerwin (Figure 3.21) on November 11, 1965. The persistence and technical excellence of the Hamilton life support personnel resulted in them gaining their reward. The backpack testing (Figure 3.22) met all requirements. Apart from minor changes, this was the first backpack used on the Moon.

Figure 3.21. Preparing for a Vacuum Chamber test
(courtesy UTC Aerospace Systems)

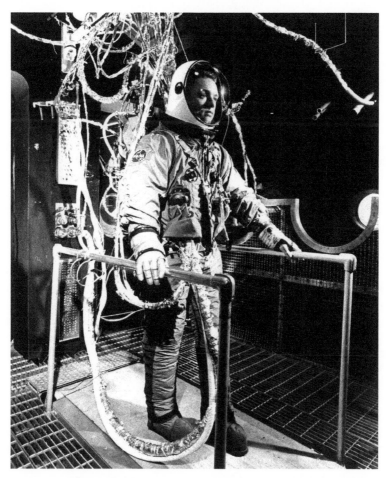

Figure 3.22. Liquid-Cooled PLSS manned test
(courtesy UTC Aerospace Systems)

Competing Suit Developments and the Pivotal Block II Competition

In Block II, NASA provided Hamilton Standard with the opportunity to mark two major milestones. By December 1964 Hamilton had to provide a pressure suit design recommendation for the A5H training fleet. As NASA did not expect A5H to be the best and final suit design before having to start lunar operations training, the decision was made to conduct a Block II pressure suit competition in June 1965. However, Hamilton was tasked with assuring A5H would be compliant to all requirements and be the best possible design. Thus, Hamilton scheduled comparative testing of all possible Apollo pressure designs in November 1964.

In the fall of 1964 there was a new dynamic in what was transitioning into the Apollo Block II spacesuit program. Hamilton's program was under new management. Breeding made a point of leaving his office door open so he could keep track of the Hamilton Apollo Space Suit Program area. He established the expectation that everyone would be at their desks and productively working at 7:30 a.m., the starting time. If an employee came in late, Breeding would call them into his office for a reprimand.

One morning, Mike Marroni came in a couple of hours late. Breeding spotted him, called him in, and demanded to know why he was not on time. Marroni explained there were problems in the sewing room during the night. He had to come in to resolve the issues and stayed until early morning. He went home, got a few hours of sleep, and came back into work. Breeding told him that was no excuse and that he should find someone to cover so he would be at work when expected. Breeding went on to say how he led by example and expected nothing less from his subordinates. Marroni then spoke with his boss, Mark Baker. Baker told Marroni that there was no one to cover. This gave Marroni an idea. He looked up Breeding's home number and started putting Breeding down as the night-shift contact on all subsequent suit lab paperwork. Every morning Marroni came in early to stand outside Breeding's office. Every time Breeding looked up at Marroni, Marroni smiled and waved. Precisely at 7:29 and 30 seconds, Marroni would scramble to his desk to immediately look productive. It took but a few days before Breeding called Marroni into his office upon his arrival. Breeding had been awakened from a sound sleep in the middle of the night by someone in the lab. A very displeased Breeding asked why. Marroni responded "You told me to find coverage and that you led by example. I was happy to cover my own hardware, but there was no one to cover. I just wanted to get a little sleep." Breeding scowled and for a long time there was silence. Then, Breeding barked "OK, this time I will make an exception but don't you abuse it! Now get out!"

Another major change was that ILC, the two Hamilton groups, and B.F. Goodrich were all working on Apollo pressure suit development in parallel. ILC was under NASA funding. The remainder Hamilton was paying for, since NASA blamed them for the problems of the program.

The first new prototype suit came from B.F. Goodrich. It was delivered in the first week of September 1964. Goodrich named it the "Mobility Suit." This was supposed to be an example of Goodrich's latest and best pressure suit technologies. The suit was a simple pressure garment with olive outer fabrics based on Goodrich's Mercury suit technology. It quickly became obvious that there had been a communications error between Hamilton and Goodrich. While Mercury-type suits were the latest Goodrich developments, they were not suitable for spacewalks. Hamilton clarified the requirements and Goodrich produced yet another prototype in a couple of weeks using Hamilton technology.

During Goodrich visits to Hamilton's Connecticut facilities, Hamilton had shared its internal developments. The Goodrich personnel saw potential in the Teflon cord, Teflon ferrule, multidirectional joint technology developed by the Mark Baker subgroup. Hamilton provided materials that Goodrich engineers took back to Akron, Ohio. This second Goodrich prototype started with an obsolete pressure suit to reduce cost and eliminate many hard details that could take months to obtain. The shoulders were retrofitted with Hamilton-type multidirectional mobility systems and Goodrich designed a tucked fabric elbow with pressure-sealing bearings at the biceps and forearm. This combination (Figure 3.23) produced a configuration that met the shoulder width requirements of Apollo and had the best-yet upper-torso mobility that probably exceeded the mobility requirements of Apollo. But how was this achieved?

Figure 3.23. Goodrich-Hamilton (second) "Mobility Suit"
(courtesy UTC Aerospace Systems)

The primary reason for this best-yet performance was the combination of the new shoulder and the addition of a bearing between the shoulder and the elbow. There had been previous multidirectional shoulders on ILC Apollo and other suits. However, the Teflon cord, Teflon ferrule joint allowed the shoulder to move up and down or rotate around the torso with less effort than any previous design. It also allowed the arm angle to move in a greater range relative to the torso with less effort than anything before. An additional plus was the shoulder tended to stay in position rather than causing the user to have to "fight" the garment to hold a position. The elbow was in some ways similar to the ILC Apollo design in that it used convolutes and two cable restraints, one on each side of the convolute. The convolute could "swing" back and forth on the side restraints allowing movement, but only in one direction. A relatively minor difference was that ILC's convolute was a molded rubber design and Goodrich's was a fabric joint that required a little less effort to bend. The significant difference was that in the preceding Apollo suits ILC had sought the optimum angle in relation to the shoulder for attaching the two restraints. The upper-arm bearing allowed the elbow to rotate such that the forearm and hand could easily move in any desired direction.

Unfortunately, the concept was firmly rejected by Radnofsky based on concerns about hard contact from arm bearings causing discomfort or injury as well as crewmember safety if a pressure seal was damaged. Given NASA's reaction, Hamilton and Goodrich focused their future development efforts on multidirectional concepts without arm bearings. However, an element within the Hamilton suit group thought rejection of the second Mobility Suit prototype was short-sighted. The prototype involved sporadic "spare-time" development of the lower torso.

November saw an evaluation flurry of pressure suit developments from Goodrich, ILC, and Hamilton with a comparative evaluation of a David Clark Gemini suit. Goodrich delivered two prototypes. Hamilton produced one suit. ILC's development support effort outshined the output of Goodrich or Hamilton with three prototypes.

The first ILC prototype to arrive was called the Playsuit. It met the Apollo shoulder width requirement and the arms had good fit in the down position, however, this prototype did not provide the needed mobility or reach. The suit was immediately returned to ILC for retrofit to an ILC design shoulder (Figure 3.24). By November 25, 1964, the Playsuit had returned to Hamilton. The evaluation showed promise.

The second ILC prototype featured the latest ILC thigh and brief section design (Figure 3.25, U.S. Patent 3,699,589, George Durney inventor). This design was not yet fully refined and had significant user comfort issues. It quickly gained the nickname "the crotch cutter" by some of the test subjects. However, this nickname should not be taken to mean the design did not have merit. This was a forerunner of the brief system used on the competition-winning ILC suit, which with some minor improvements saw service on the Moon and on Skylab.

Figure 3.24. The ILC Playsuit
(courtesy ILC-Dover LP)*

The third ILC prototype featured Goodrich-style shoulders. The purpose of this prototype was to demonstrate that ILC was still capable of pressure suit manufacture should NASA select this Goodrich shoulder for the A5H training suit design. This garment was sent to Hamilton. The similarity of performance in comparative testing to the Goodrich suit with the same design shoulder showed that ILC could produce the Goodrich shoulder. This was the last complete prototype pressure suit from ILC under the Hamilton Apollo contract.

November also saw the debut of Hamilton's first complete design and prototype manufacture called "the Tiger Suit." This was an interesting conformal-fitting, olive-and-white two-tone pressure garment. This prototype was the collaboration of Ed Vail and Jim Hopper. A possible reflection of the anticipated importance of this first complete Hamilton suit design and build, a

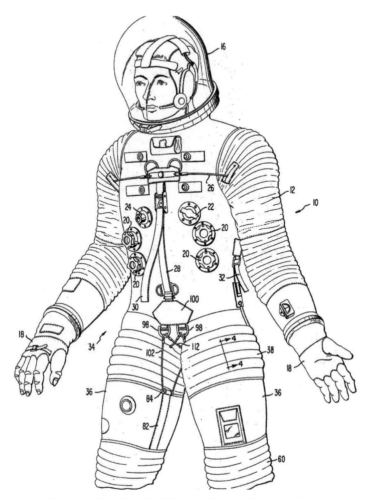

Figure 3.25. The ILC/George Durney Walking Brief
(courtesy U.S. Patent Office).

photograph of the principal actors was taken (Figure 3.26). This shows, from left to right, Dr. Edwin Vail (Head of Space Pressure Suit Development), Edward Marshall (Vice President and Head of Space & Life Systems), Hamiltonaut Ed Brisson in the not-yet-completed garment, Edmond Garraventa (Hamilton Lunar Module Program Manager), and William E. "Bill" Diefenderfer (President of Hamilton).

The Tiger Suit did not have a pressure gage. It was assumed that the pressure of the ventilation air entering the suit from the test rig would also be the pressure inside the suit. This would have been true had not a blockage in the suit's inlet caused a significant pressure drop before the ventilation gas entered the suit. Initial evaluations showed the suit to have break-through mobility, which was immediately reported to Hamilton management who, without

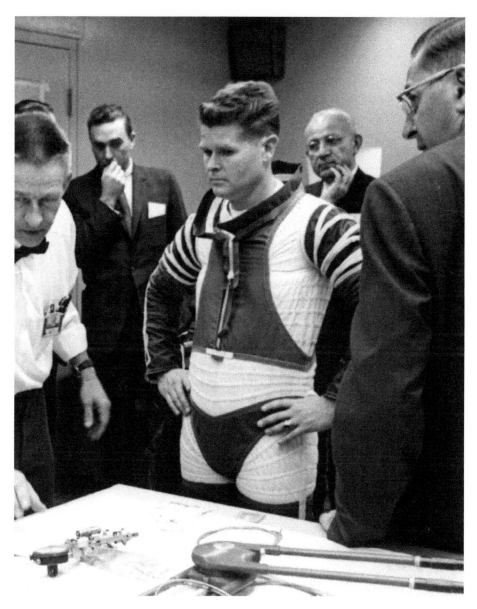

Figure 3.26. The Hamilton "Tiger Suit" in precompletion review
(courtesy UTC Aerospace Systems).

hesitation, informed NASA. NASA personnel straightaway arranged travel to Windsor Locks for a demonstration. NASA requested to have the test start with the suit initially pressurized and the suit subject lying on his back. The subject would then rise to his feet, something no previous suit could do. As Hamilton management expected this could be easily accomplished, they agreed.

In parallel, Tiger Suit testing continued. After one of the suit tests, Andy Hoffman, the Project Engineer, noticed a red mark on Ed Brisson's stomach. This was caused by the suit's inlet duct becoming excessively hot, which was the result of the blockage. Once corrected, the effort required for movement was higher than that of the ILC Apollo configurations. Hoffman reported this to his superiors urging the demonstration be cancelled. However, Hamilton management felt it was too late to cancel and Brisson was challenged to perform the demonstration as planned.

Brisson repeatedly practiced with the Tiger Suit in preparation for the demonstration. On the designated day the demonstration started with Brisson lying on a table. The suit was then inflated. Try as he might, Brisson was unable to rise. Finally helped to his feet, Brisson continued the demonstration. Unseen, a shoulder cable was wearing under repeated friction. When the cable failed in front of NASA and Hamilton management, the left arm grew in length and became unbendable. The demonstration was subsequently terminated.

Apollo testing in November 1964 involved NASA providing a David Clark Gemini training suit to Hamilton for comparative mobility testing (Figure 3.27). A David Clark Company representative supported the evaluation. The comparative test results demonstrated that Apollo suit technologies were finally surpassing the Gemini benchmark.

The winner of the November 1964 testing was a Goodrich design called the XN-20. Analysis of the testing showed the design met all upper-torso requirements and was the best Apollo suit design yet. This would have been the basis of the A5H training suits had it not been for ILC being confident that they could provide better designs for testing by the end of December 1964. ILC's vision was to incorporate the best pressure suit ideas into one garment. ILC named the subsequent prototype the "Composite Mockup" suit.

As a result, Hamilton requested and NASA accepted extension of the pressure suit design selection deadline to January 15, 1965. As more time was provided, the extension spawned three design efforts, one by ILC and two by Hamilton. Hamilton's primary effort involved "Doc" Vail and Jim Hopper redesigning the "Tiger Suit" and making it a competition winner.

The second Hamilton effort stemmed from Vail having so much confidence in Hopper that he effectively made Hopper the creative center of Hamilton's mainline efforts earlier in 1964. However, this left a pool of talent that had been placed at Vail's disposal underutilizedpeople such as Mark Baker who was immediately under Vail in Hamilton's suit development structure, John Korabowski who had been part of the first Hamilton mobility effort that started in late 1963, and some newly recruited engineers such as Michael Marroni and Doug Getchell.

By June 1964 Baker had begun guiding a Hamilton effort parallel with that of Vail's. This subteam methodically explored every conceivable elbow, knee, waist, and brief mobility idea by making prototypes of each section, and then quantitatively measuring them for a comparative database. This was supported by around-the-clock master seamstresses who were drafted from Hamilton's

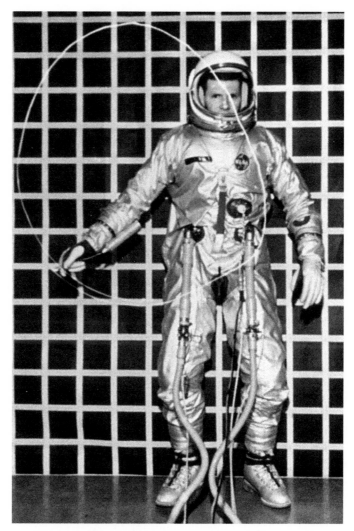

Figure 3.27. Gemini Evaluation Suit
(courtesy UTC Aerospace Systems)

fiberglass aircraft parts manufacturing. These women were so skilled that they could from verbal descriptions or rudimentary sketches make the patterns and start assembly of the mobility element suit section. At shift change the women stayed to explain what needed to be done to assure the prototype section would be completed by morning. In the morning the engineers came in and readied the prototypes for test and then ran the tests to produce quantitative data that permitted unbiased comparisons.

By September 1964 the Baker effort resulted in a Teflon cord and Teflon ferrule concept that Goodrich used in the multidirectional shoulders of their second Mobility Suit (Figure 3.23). Further development produced what appeared to be competitive designs for the various suit areas. However, by mid-December 1964 these elements had not been incorporated into suits for manned testing. Extension to the end of December 1964 gave the Baker group the opportunity to compete the manufacture of suits for manned evaluations.

To expedite developments to meet the schedule deadline, Baker used two leftover pressure suits. However, to be able to repeatedly reuse the same pressure suit with new prototype mobility elements required sewing skills that were nothing short of wizardry. If the seamstresses simply took out an old seam and sewed a new mobility element onto an existing torso section, the fabric of the existing torso section would be weakened as a result of the sewing machine needle making additional new holes in the fabric. The first or second replacement would cause the existing torso section to tear when the suit was pressurized. Sewing sections together by hand would not only be too slow but the thread tension would be uneven, which would cause the joint to fail under pressure load. To be able to reuse the same pressure suit the seamstress had to sew the new section onto the pressure suit making sure the sewing machine needle was passing through existing holes in the pressure suit fabric. This allowed incremental refinements in designs. To speed development, one suit was used exclusively for upper-torso features while the other suit was modified to permit lower-torso development and testing.

After eight months the Vail and Hopper effort had proven unsuccessful and ILC had not yet provided their Composite Mockup suit. However, the Baker effort had literally progressed around the clock through the month and into the plant's Christmas and New Year shutdown. On December 28, 1964 the Upper Torso Suit (Figures 3.28 and 3.29) was under test at Hamilton. The Upper Torso Suit design was based on the second Goodrich Mobility Suit with an improved shoulder but the upper-arm bearing and fabric convolute were replaced with a Teflon cord, Teflon ferrule multidirectional elbow that had the same range of motion but required more effort to move, although not to the level of causing discomfort or bruising. The Hamilton "Thigh Suit" followed a few days later.

With the January 15th deadline looming for the A5H design recommendation, ILC requested an additional month delay. As NASA Headquarters in Washington had been extremely displeased at the previous one-month delay, NASA-Houston made it abundantly clear to Hamilton that there would be no more delays. On January 12 and 13, 1965, a Friday and Saturday, Hamilton presented the comparative test results from the preceding two months to aid selection of the best design to support A5H training suit production.

The best design was effectively a tie between the last B.F. Goodrich configuration called the XN-20 and the combined features of the Hamilton Upper Torso Suit and Thigh Suit. Hamilton demonstrated the Upper Torso Suit and XN-20 suits to explain the differences. Hamilton recommended the Upper Torso Suit and Thigh Suit systems to be the basis of the A5H training suit design. NASA accepted the recommendation. The Hamilton Apollo program

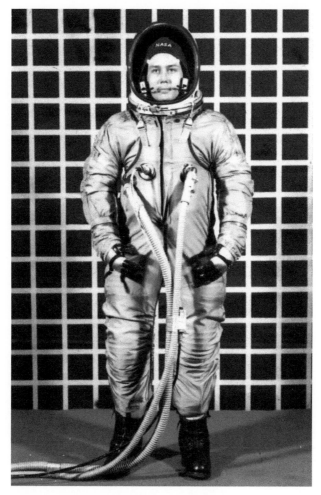

Figure 3.28. Upper-Torso Suit without Cover Garments
(courtesy UTC Aerospace Systems)

continued funding for ILC's Composite Mockup. While this was too late for A5H production consideration, it was a potential candidate for the then-expected Hamilton-ILC 1965 Block II Competition prototype. To complicate matters, ILC's management wanted to rerun the Hamilton-run November/ December suit testing at ILC. When that was denied, ILC declared that it would not manufacture a Hamilton design.

To oversee Apollo spacesuit developments, NASA had deployed a NASA engineer, Bob Grafe, to Windsor Locks, Connecticut in the summer of 1964. In his January 22, 1965 report, Grafe reported on the arrival of what was expected to be a revolutionary new prototype from ILC, the Composite Mockup Suit. What was received took both Hamilton and most of NASA by surprise. To

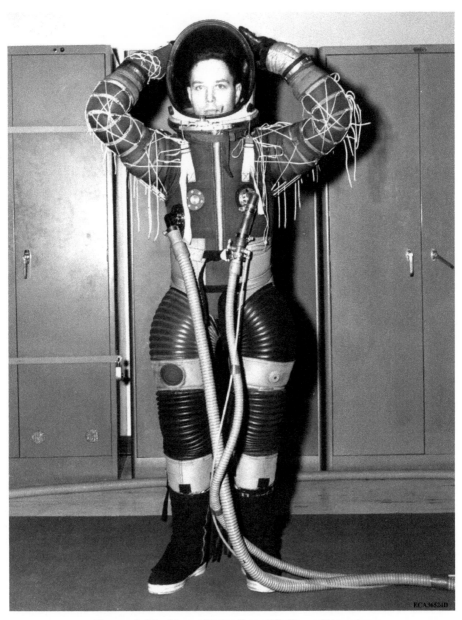

Figure 3.29. Upper-Torso Suit with Cover Garments
(courtesy UTC Aerospace Systems)

quote Grafe's report "ILC has shown a complete lack of talent in mobility development thus far. The composite mockup, which was delivered to Hamilton this week, seemed to be the finale of ILC inadequacy. The mockup consisted of only the torso and legs, the arms and shoulders were to be attached by

Hamilton. The legs can be salvaged from this suit but the torso is unusable." Hamilton issued a "stop work" order to ILC on all Apollo development activities. Diefenderfer directed Hamilton personnel to explore, with all haste, developing the option of making Goodrich the Apollo pressure suit supplier. This was not shared with ILC, probably to preserve the option of continuing with ILC should NASA not approve.

Attempting to understand the events of the Apollo Space Suit Program from November 1964 through the summer of 1965 is very much like assembling a puzzle with some of the parts missing. The incomplete "Composite Mockup" suit was certainly not ILC's vision of the best pressure suit ideas incorporated into one garment. It embodied no new ideas. From the descriptions, it appears to have been a reject torso assembly from the A4H production line. However, by the first week of February 1965, ILC appears to have completed an all-new revolutionary design Apollo prototype that did incorporate ILC's vision of the best available pressure suit ideas into one garment. With David Clark Gemini-style rear entry and other unique features from other organizations, the prototype could not be created by modifying an existing pressure garment. The unique metallic prototype had to be designed and then manufactured, which could easily take two months under a highly accelerated schedule. ILC-Dover probably started that design process no later than the start of December 1964. Given the timing, this prototype certainly started off as the Composite Mockup suit. However, the time associated with designing, ordering, and making the unique parts required meant the goal of being ready by the end of December 1964 as promised by ILC's New York management was not possible. ILC's management was very displeased that Hamilton and NASA did not wait for ILC and selected a non-ILC design for the Apollo A5H suit. What lied behind the decision to deliver an incomplete, mismanufactured torso for the Apollo program and call it the Composite Mockup? This will likely always remain a mystery.

What is known is that ILC completed a "State-of-the-Art Suit" rear-entry prototype (Figure 3.30) by the start of February 1965. This design embodied the best B.F. Goodrich, David Clark, Hamilton, and ILC concepts to produce a revolutionary prototype, which was a milestone pressure suit and the basis for the first suits used on the Moon. For reasons that have been lost, International Latex management elected not to share this key advancement with Hamilton and top NASA management.

By the beginning of February 1965 ILC had agreed to manufacture Hamilton designs and found itself once again funded for Apollo pressure suit development. While both ILC and Hamilton were proclaiming their commitment to working together, events in motion were to the contrary. Hamilton's pursuit of the Goodrich option for Apollo suit manufacture did not desist and ILC never demonstrated this or any other Apollo suit developments to Hamilton.

In mid-February 1965 NASA engineer James "Jim" O'Kane started making trips to the ILC-Dover plant to coordinate development of the next NASA

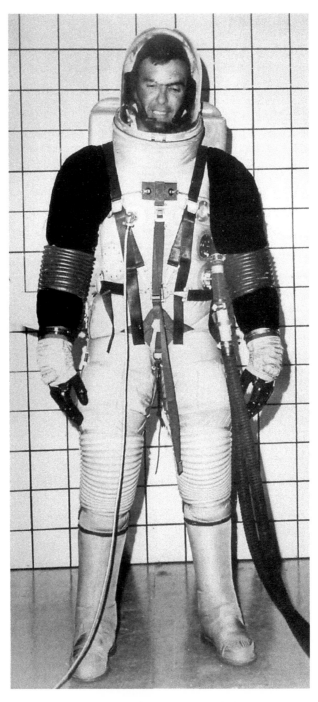

Figure 3.30. ILC Rear-Entry SOA Suit with "blacked out" shoulders
(courtesy ILC-Dover LP)*

O'Kane-Jones polycarbonate full-bubble helmet to specifically fit the new ILC rear-entry prototype (Figure 3.30).

What is interesting is the ILC-Dover facility did not make any effort to conceal their collaboration with O'Kane. Quite the opposite, ILC-Dover supported local paper coverage of the activities. As the suit did not have cover garments over the shoulder and elbow sections, ILC blacked out the details of the mobility systems in pictures released to the public (Figure 3.30). It is not known whether Len Shepard, Art Finklestein, or others at the Dover facility were hoping that Hamilton would pick up on ILC's latest work. If they had picked up on it, the attempt was unsuccessful. What is known is that this suit was so externally similar to ILC's Block II competition prototype of June 1965 that for many years spacesuit historians mistook photos of this earlier rear-entry prototype for those of the ILC suit that won the Apollo Block II competition.

In early February 1965 Hamilton supported the creation of a NASA public relations movie. This created an additional program challenge as it competed with A4H deliveries, A5H development, and preparation for the Apollo Block II competition.

Hamilton took the ILC Composite Mockup and retrofitted it with A5H-style shoulders and arms. As this was the third use of the Composite Mockup name in Apollo, Hamilton renamed this the "Mobility Suit" (Figure 3.31). Because the Hamilton Mobility Suit was immediately consumed by service in NASA vehicle development and cinematic activities, Hamilton subsequently modified another A4H-like test suit that was subsequently called the "A4H Retrofit Suit." The ability to upgrade existing suits to meet Apollo width and mobility requirements sufficiently pleased NASA that Hamilton was directed to retrofit two late A4H training suits with Hamilton arms and shoulders.

After at least two weeks of telephone dialog between Hamilton's President William E. Diefenderfer and the Chief of NASA's Crew Systems Division Dick Johnston concerning ILC, the men met face to face on March 3, 1965 at the Manned Spacecraft Center in Houston. ILC representatives were not invited. Diefenderfer was accompanied by Hamilton's Vice President of Space and Life Systems Edward E. Marshall. Marshall explained why Hamilton felt ILC was not capable of supporting the Apollo program and that Hamilton needed to have a contractor that would, without question, support Hamilton in the Apollo Block II Pressure Suit competition that was to start in just 14 weeks. Apparently, the people at the meeting were not aware of ILC's latest prototype. At the conclusion of the meeting, Hamilton was given written authorization to terminate the ILC Apollo contract and enter into contract with Goodrich. ILC remained under contract to Hamilton for completing A4H training suit deliveries. Hamilton urged that ILC should not be funded for any additional Apollo development or for a Block II competition prototype. Johnston assured Diefenderfer that NASA would not fund an ILC competition prototype. It is unclear who else in the Crew Systems Division had been privy to these Hamilton/NASA discussions prior to this meeting. The number of NASA participants

Figure 3.31. Hamilton Standard "Mobility Suit"
(courtesy UTC Aerospace Systems)

may have been limited to maintain an ILC pressure suit continuation option until the outcome of the meeting.

After the meeting, Marshall called Hamilton's Purchasing Manager William F. Gerety who sent a predrafted telegram to Dr. Nisson A. Finklestein, the then-Vice President and General Manager of ILC's Government and Industrial Division, informing them of the contract termination and requesting an immediate reply telegram confirming receipt of the notification. Waiting in Akron were Hamilton representatives ready to sign a contract with Goodrich. The day ended with no reply from ILC. At 9:25 a.m. on March 4, 1965 Gerety received the confirmation telegram from ILC's Art Finklestein. Gerety then called Akron, the contract between Hamilton and Goodrich was signed, and the Apollo spacesuit program had a new pressure suit supplier. Hamilton engineers remained in Akron to facilitate interactions between Hamilton and Goodrich. Goodrich immediately started work on A5H activities in Akron. Within days, Goodrich personnel were also at Hamilton's facility in Connecticut thereby creating the capacity needed to aid Apollo development from two sites simultaneously.

ILC's management was caught by surprise. ILC-Dover personnel felt betrayed by Hamilton Standard, a depiction probably encouraged by ILC's New York and Dover management. This Hamilton move certainly generated long-lasting resentment and distrust of Hamilton by many ILC-Dover personnel. This was in contrast with the preceding three years which had resulted in a few ILC-Hamilton friendships between technical personnel.

On March 18, 1965, Cosmonaut Aleksey Leonov made the world's first spacewalk (Figure 3.32). NASA mobilized for a U.S. response in the form of the heroic effort that took the U.S. from not having a suit system capable of use in a vacuum to the first American spacewalk in 11 weeks. In support of the Gemini effort the David Clark Company developed the first U.S. space-certified thermal overgarments to protect astronauts while spacewalking. From March to June 1965, while NASA was focused on Gemini, a sense of calm and normalcy seemed to settle upon the Hamilton Apollo program.

Astronaut Mike Collins was to be the suit evaluator in the Block II suit competition that was starting on June 15, 1965. The first prototype A5H, later called AX5H, was to be sized to Collins as it was to additionally serve as the Hamilton/Goodrich competition prototype (Figure 3.33). The second A5H prototype in, being manufactured at the same time, was being sized for Dan Galvin, a Hamilton suit evaluator. The Collins and Galvin prototypes were completed and delivered in parallel with continuing mobility system development. The Collins suit was sent for use in a Command Module critical design review. Goodrich had developed a pressure-sealing zipper that showed promise in dramatically reducing suit leakage. The Galvin suit was sent to Goodrich for the addition of this feature.

In April 1965 NASA desired a smaller neck ring and more compact helmet with greater range of visibility than the late A4H and the A5H suits. This became a requirement for the A6H flight model. The A5H training suit configuration at this point was not going to be changed to permit retaining the delivery schedule on the prototypes and subsequent production.

NASA additionally desired these new features be part of the Apollo Pressure Suit competition that was to start in June 1965. This necessitated a redesign of the thermal micrometeoroid outer-garment, the helmet, visor assembly, suit-side helmet disconnect, and neck area torso in tandem with the creation of new tooling, the manufacture of new suit components, and assembly of the spacesuit prototype (less backpack). Hamilton, being eager to please NASA, accepted the challenge. As a result of schedule constraints, the new Hamilton-Goodrich com-

Figure 3.32. The world's first human step into space
(courtesy NASA)

petition suit was built at Hamilton. Operations literally ran 24 hours a day, 7days a week.

In parallel, the Galvin AX5H suit was completed by Goodrich and included Goodrich's pressure-sealing zipper. The Galvin suit showed no measurable leakage in a test of its new entry system, which was a remarkable feat.

In 1964 NASA personnel Jim O'Kane and Bob Jones proactively initiated development of a NASA polycarbonate full-bubble helmet. In May 1965

Figure 3.33. The first AX5H suit
(courtesy UTC Aerospace Systems)

O'Kane requested an opportunity to present to Hamilton Apollo spacesuit personnel their latest prototype (Figure 3.34) while he was in the area. As O'Kane
insisted that all program personnel be present the only opportunity was on a
Saturday. While NASA's prototype helmet did not provide the required optical
quality, it did demonstrate the simplicity of a polycarbonate bubble attached
directly to the neck ring. The NASA prototype included an easily removable
visor assembly (Figure 3.35). In this presentation O'Kane campaigned for
Hamilton to internally fund the development of a helmet that would be smaller

Figure 3.34. NASA's 1965 Polycarbonate Full-Bubble Helmet
(courtesy UTC Aerospace Systems)

than the Apollo requirements, or as a minimum for Hamilton to put pressure on NASA management for a smaller helmet.

Their prototype involved Jones and O'Kane measuring the Apollo astronauts and allowing half an inch extra in any direction to create their more compact geometry. Members of the Hamilton audience questioned the possibility of a future astronaut having a larger head than the present cadre and pointed out

Figure 3.35. O'Kane/Jones 1965 Apollo Visor concept
(courtesy UTC Aerospace Systems)

that the Jones/O'Kane approach violated the Apollo contract, which called for the development of helmets to support male heads up to the 95th percentile. O'Kane was confident that their sampling was adequate and that the Apollo requirements would change.

While O'Kane left without gaining Hamilton support for all his requests, he was partially successful. Hamilton immediately internally funded two lines of helmet development. One was based on the current A5H helmet but with a larger field of vision. The other was a polycarbonate full-bubble helmet made to contract sizing at the 95th percentile. A5H-based development was supported by Ken Griffin, the engineer who developed the C3 helmet with the first optical quality polycarbonate visor. Griffin's efforts resulted in "half-bubble" and "three-quarter bubble" prototypes that met the optical quality standards and featured shells that did not obstruct vision beyond that caused by a sun visor assembly. The first Hamilton full-bubble attempts were performed by James Hopper. While Hopper quickly produced a prototype, his efforts never reached optical quality. In late summer Hamilton continued the Full Bubble Helmet effort but this was turned over to Ken Griffin.

By mid-May both AX5H prototypes were in remanufacture to the latest mobility developments. Also in May a third AX5H-type suit was under construction at Hamilton. This was being custom-sized for NASA engineer, suit evaluator, and Pressure Suit Manager Jerry Goodman.

The introduction of extravehicular thermal overgarments in the Gemini program provided an opportunity for performance comparisons. The Gemini outer covers developed by David Clark were proving more durable. In parallel, the feasibility of being able to design an overgarment that could meet Apollo impingement requirements had become a topic of serious discussion. The credibility of meteor and meteorite strikes during a lunar mission was coming under question by NASA, Hamilton, and ILC personnel alike.

It was probably not a surprise that sample testing still did not meet Apollo impingement requirements. Fortunately for the program a risk analysis at the time indicated that the likelihood of an asteroid strike, the driver for the requirement, was significantly less than a solar flare during a lunar mission. NASA had already accepted that the solar flare risk was sufficiently remote that incorporating high-level radiation protection into Apollo vehicles and spacesuits was not required. The combined data and analysis helped NASA decide that the Apollo impingement requirements were too conservative to be viable. Gemini standards were subsequently adopted. NASA had funded the development of the Gemini overgarments. Since both Hamilton Standard and David Clark Company were under contract to NASA, it seemed a simple matter to NASA for a transfer of technology from Gemini to Apollo. Nevertheless, there were issues in having the David Clark Company aid the Hamilton program.

Most David Clark Company personnel felt that Hamilton had discarded them too quickly at the start of the Apollo program and hard feelings lingered. Additionally, Hamilton was now a competitor. However, the David Clark Company was a loyal supporter of the U.S. space program and subsequently aided Hamilton's Apollo thermal garment development.

The Apollo Block II Pressure Suit Assembly competition was for a complete pressure suit system that addressed the needs of launch/reentry and extravehicular activity by donning accessories. The competition was managed by Dr. Robert "Bob" Jones. As NASA's primary suit evaluator was Astronaut Mike

Collins and the Apollo program planned for custom-made pressure suits, the competition prototypes were to be sized to Collins. As Jack Mays and Collins were of similar build, Mays was selected as the second suit subject for the competition. He is a native of San Antonio, Texas and served four years in the Navy where he gained pressure suit experience. He returned to San Antonio to work at Kelly Air Force Base. He recalls, "A friend of mine at Kelly told me NASA was looking for pressure suit people, next thing I knew NASA told me to come to Houston for a talk. I came back to SA packed up and moved to Houston. That was in August of 1962." This positioned Mays to be a NASA technician and suit subject on the Apollo program.

All prototypes were to be delivered to NASA in Houston by noon June 15, 1965. The first prototype to arrive at the competition was the David Clark suit (Figure 3.36, left).

This suit was a significant departure from their rear-entry Gemini (Figures 1.19 and 4.1) and Apollo Block I suits (Figure 3.36, left). While all these designs used David Clark's Link-netTM restraint system, the AX1C had shoulder bearings and other enhanced arm mobility features plus a front-top entry like the 1962-64 Apollo suits. However, the David Clark suit and the

Figure 3.36. Jack Mays and the Block II Competition Suits
(courtesy NASA)

competing Hamilton-Goodrich suit both did away with cable attachments to the neck ring. This allowed unassisted donning and doffing. The David Clark suit additionally featured a polycarbonate full-bubble helmet that met all Apollo requirements except optical quality. It also had accommodations for a liquid cooling garment. David Clark elected to champion the concept of an "integrated thermal meteoroid garment" that attached to the torso and gloves, rather than separately donned cover garments as specified on the Apollo contract. This simplified going for a spacewalk as the astronaut only had to don a visor assembly, overboots, and a backpack before setting foot on the Moon.

The other suit at the beginning of the competition was the Hamilton-Goodrich suit (Figure 3.36, right). Even though the David Clark Company was Hamilton's competitor, it provided glove finger lights and technical support on the thermal overgarments of the Hamilton-Goodrich competition prototype. As a result of time constraints imposed by NASA's request for a smaller neck ring and a higher visibility helmet, Hamilton was never able to test its prototype as a complete spacesuit before delivery to the competition. Hamilton's schedule challenges were additionally increased by the only opportunity for a Collins predelivery fitcheck at Hamilton being June 12, 1965, just three days before the required delivery date. Hamilton requested an extension, which NASA denied. To avoid disqualification, Hamilton hired a private jet to make the competition deadline. Moreover, schedule compression resulted in the competition suit leaving Windsor Locks unfinished. Andy Hoffman, one of the supporting engineers, still remembers vividly the final lacing and assembly of the suit in flight.

The time impacts of accepting the A6H redesign for the Hamilton competition prototype quickly became apparent. The jacket part of the Thermal Meteoroid Garment did not fit correctly, causing the visor assembly to constantly detach from the helmet. Further into the evaluations, the helmet shell blew off its neck ring causing the shell and visor portion of the helmet to rocket into the air during a ladder-climbing exercise. While no one was seriously hurt, this decompressed the suit and could have caused serious injury. By design, this should have been impossible. However, the failure investigation revealed that not only had the neck ring of the helmet not been properly bonded to the helmet's fiberglass shell but the metallic details that would mechanically hold the shell to the neck ring during a bond failure were missing. Both errors were due to the work being performed by night-shift personnel who were unfamiliar with spacesuit work and were lacking engineering supervision.

While the AX5H Galvin suit that had demonstrated no measurable leakage under pressure had been delivered to Hamilton, Hamilton elected not to send it to the competition under the assumption that it was unnecessary and would place an additional burden on the onsite government inspectors who had to process other Hamilton-Goodrich items.

The generally anticipated Block II competitors were Hamilton and the David Clark Company, but ILC had been attempting to make the competition. Chief of NASA's Crew Systems Division Dick Johnston had been true to his decision to not fund an ILC competition suit. However, ILC's New York headquarters was confident that NASA would relent and thus initially refused to fund the ILC competition suit. The ILC-Dover management and personnel were caught in the middle. What consequently transpired was a monument to resourcefulness.

The ILC-Dover team recognized they were behind the competition both in terms of time needed to produce a competition suit and customer perception of ILC capabilities. To counter the perception, ILC-Dover management wanted it acknowledged that their additional Apollo suit development was on par with Hamilton and ahead of David Clark. Expecting that the Hamilton entrant would be an AX5H, ILC-Dover representatives made a presentation to NASA's Charles Lutz advocating that two preceding ILC nonprogram configurations had been significant to the Apollo effort and should be provided NASA designations retroactively. The presentation was successful in earning ILC's competition suit the Apollo AX5L designation. However, the creation and delivery of the ILC competition suit was uncertain.

ILC's New York headquarters did not authorize internal funding for an ILC competition suit. ILC's President Wally Heinz was convinced that NASA would recant its position and fund ILC to be in the competition. As the needed start time grew near and then passed, neither NASA nor ILC headquarters revised their positions. All would have been lost if ILC-Dover personnel had not proactively started building their competition prototype during after-hours and in their spare time. However, volunteerism has its limits. In World War II and the Korean War, there was an American military term called "midnight requisition" to overcome lack of needed support from normal supply channels. Most, if not all, of the Dover personnel were military veterans. There are stories of ILC personnel climbing over crib fencing and breaking into locked rooms at night to obtain needed materials. Perhaps one of the reasons ILC's Rear Entry State of the Art Suit so completely disappeared from history is that it contributed metal parts with a long lead time that could not have been obtained elsewhere given the time and monetary constraints. Regretfully, even resourcefulness has its limits. The dipping facility needed to create the molded convolutes was controlled by another ILC division. Making the convolute joints required formal funding. Only after it was too late to make the competition start date did ILC headquarters finally provide internal funding.

NASA allowed ILC to enter the competition two weeks late with their AX5L prototype. Other than two features the exact differences between the ILC suit (Figure 3.36, center) and ILC's earlier Rear Entry State of the Art Suit (Figure 3.30) are unknown. One difference was the color of the rubber in the elbow convolutes. The competition suit featured a blue rubber blended to match the outer fabric of the garment. The other known difference was the introduction of a Goodrich pressure-sealing zipper in the Gemini-style rear-entry system. This gave the AX5L the lowest leakage rate of the competing suits tested. One strategic feature of both ILC suits was the walking brief and thigh restraint system (U.S. Patent 3,699,589, George P. Durney inventor). This permitted walking with significantly lower effort than any preceding design.

The ILC competition suit also featured gloves with improved, steel cable, multidirectional wrist joints. This last addition was the culmination of a glove effort by ILC's Dixie Rinehart. His latest design would not only support the early Apollo missions but, with minor derivations, the last Apollo missions, Skylab, and Apollo-Soyuz as well. As a result of a twist of fate the space service of Rinehart's design would not end there.

The ILC competition suit utilized the latest NASA polycarbonate full-bubble helmet and visor system developed by Jim O'Kane and Bob Jones (Figure 3.36, on table).

Entry of the ILC suit into the Block II suit competition encouraged Hamilton to offer two prototypes for NASA consideration that Hamilton and Goodrich felt would make a difference to the NASA decision. NASA declined evaluating both.

As part of the Block II evaluations the contractor teams were permitted one 24-hour opportunity to service their suits. Hamilton spared no effort to maximize the performance of their suit. Since the contractor teams were not allowed in NASA buildings after normal business hours, Hamilton turned the adjacent parking lot into a manufacturing area. Fortunately, the weather was obliging. Hamilton personnel worked through the night using every minute of the 24 hours.

NASA unofficially informed Hamilton at the end of July 1965 that ILC would be the Apollo Block II Pressure Suit Assembly provider and that NASA was assuming the role of Block II management and integration. In just nine months, Hamilton had gone from NASA's Apollo spacesuit provider to just the Block II backpack provider and even that was in potential jeopardy.

With the entry of the ILC competition suit, ILC made the claim that it was designed and manufactured under internal funding, thus the design was their intellectual property. Using this, ILC at least attempted to dictate the terms going forward. ILC made a presentation to NASA that they should be made the prime contractor and be allowed to fund Westinghouse for backpack development. While this was probably not credible in the eyes of NASA, given the surprises of the past month, Hamilton took the presentation seriously. Hamilton had yet to successfully demonstrate in manned vacuum chamber testing.

Subsequent cherry picking of competition events produced a variety of versions regarding this turning point in Apollo spacesuit history. Perhaps the most interesting aspect of this history relates to Radnofsky's vehement rejection of the Goodrich (Second) Mobility Suit for safety reasons relating to the use of upper-arm bearings. This caused the Hamilton Apollo suit program to devote all its resources from October 1964 to June 1965 on alternative methods of gaining the needed mobility within the required shoulder width. The ILC Competition Suit had a shoulder and arm design that was similar to the Goodrich prototype and also featured upper-arm bearings. Radnofsky had no safety issues with the upper-arm bearings on the ILC suit.

The official report that followed much later provided something to support

Figure 3.37. David Clark Thermal Meteoroid Garments in 1965
(courtesy UTC Aerospace Systems)

almost any opinion of the competition. In the helmet section, Dr. Jones declared the helmet that he and Jim O'Kane had developed to be the winner, despite it not meeting any of the Apollo requirements and having the worst optical quality of the three helmets tested. Given that, was pressure suit section testing credible? It appeared to be but many refused to believe so.

In parallel with contractual battles and changes in leadership, Hamilton was still under contract to NASA for A4H suit deliveries and finishing some development activities. The next Apollo program thermal overgarment prototypes were delivered to the Hamilton-Apollo program by the David Clark Company. The overgarments embodied improvements that included a separate visor cover and jacket rather than a parka-style overgarment. Since NASA decided to relax the thermal and impingement requirements, this allowed for the creation of thinner, more body-conforming overgarments for evaluation (Figure 3.37). The testing included a Hamilton-designed visor assembly that bore an amazing resemblance to the visor system used on Apollo 11 through 13. Since the Block II suit evaluations were performed in secrecy with the rules changing after the start of the competition, how can anyone have an informed opinion of the fairness of this pivot point in Apollo spacesuit history? Fortunately, the competition required two suit subjects to simulate mission surface activities. One was Jack Mays (Figure 3.33, center), who is known for his frankness and honesty. Mays clarified the results perfectly when he said, "All three suits were good but the ILC suit was the best." The heroic efforts of the ILC-Dover personnel paid off. The best pressure suit in the Block II competition was ILC's. With minor improvements and changes, this design would be the pressure suit used on the Apollo 11 through 14 lunar missions.

CHAPTER 4

The Short-lived Apollo Block I and David Clark Apollo Contributions

The design and manufacture of the first pressure suits used on the Moon could easily have been a story of Worcester, Massachusetts and the people of the David Clark Company. What separated this potential version of history from what is recorded in the history books was a decision or perhaps a series of decisions. Certainly, the first suits used in Apollo manned missions would have been manufactured in Worcester, but for a tragic accident. Even though this path on the Apollo journey did not reach the Moon as part of the program, elements of the ingenuity and dedication of the people of David Clark were present in humankind's first lunar explorations.

The evolution of lunar spacesuit innovation (Figure 4.1) began with the David Clark Company starting a business offshoot to design and manufacture pressure suits at the conclusion of World War II (WWII). This endeavor included partial-pressure suits that were an evolutionary derivation of the Anti-G suits produced for pilots by David Clark during WWII and full-pressure suits.

Early on in the development of David Clark full-pressure suits was a design simply named Model 18. Its designers, Joe Ruseckas and John Flagg, developed a rear-entry system for ease of donning and doffing the pressure garment.

John Flagg was a WWII Army Air Corp transport pilot having learned to fly near Worcester, Massachusetts while working as the night-shift machine operator at the David Clark Company. Flagg returned from the war and was rehired by David Clark as the first Director of a newly established Research and Development department.

Joseph "Joe" Ruseckas was a professional pilot before WWII, serving as both an instructor and a Federal flight examiner. Joe initially ferried fighters before progressing to flying transport aircraft in the China-Burma theater of WWII. After the war, Ruseckas returned to his civilian roles. One of his students was David M. Clark and a friendship resulted. When the airfield where Ruseckas worked was closed, Clark offered him a job at David Clark Company developing pressure suits for experimental, high-performance aircraft.

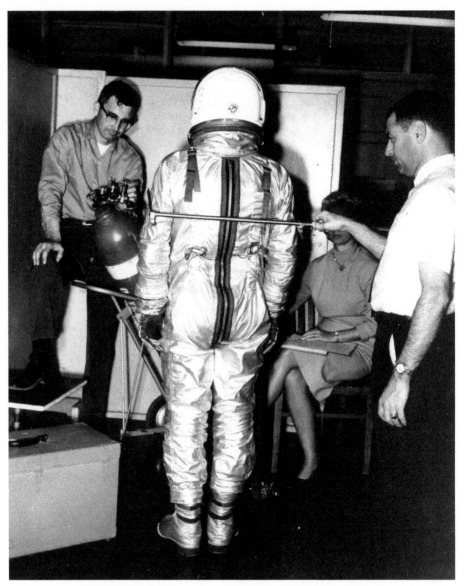

Figure 4.1. David Clark rear entry for Gemini and Apollo
(courtesy NASA)

While this rear-entry approach temporarily fell from use on subsequent David Clark models, it was destined to see an historic resurrection.

In 1957 the David Clark Company won the pressure suit contract for the X-15 program with a unique three-layer suit design. The inner layer was a conventional bladder but the middle layer was a fishnet weave of cords called Link-net™ (U.S. Patent 3,081,459, David M. Clark inventor). This provided a simple pressure layer that was free from hard details that could cause extreme discomfort or injury during launch and reentry. It also provided a limited, multidirectional capability, although at a penalty of the effort required to force the fishnet-looking cords to slide in relationship to each other. The outer layer was aluminized nylon that was donned separately and provided abrasion protection and a marginal amount of thermal protection in near-space vacuum.

Operating as an aircraft the X-15 ultimately reached Mach 6.72. Starting in 1962, 8 X-15 pilots on 13 missions traveled above the 50-mile altitude limit to qualify as spaceflights, thus making the X-15 pressure suit a spacesuit too.

In September 1961 NASA's Langley Research Center issued a contract to David Clark, among other possible suppliers, for pressure suit studies and the production of preliminary prototypes. At the end of 1961 the David Clark Company sought to be the Apollo pressure suit supplier by teaming up with another organization. The resulting David Clark prototype (Figure 1.19) supported a Hamilton Standard and David Clark proposal for the Apollo spacesuit contract. In the Apollo Space Suit Assembly proposal evaluations, NASA preferred the Hamilton backpack life support concept and the International Latex Corporation (a.k.a. ILC) pressure suit. NASA offered the contract to Hamilton only if it used ILC as the pressure suit provider. Had Hamilton during preaward negotiations succeeded in convincing NASA that its analysis of the potential field of suppliers was correct, the David Clark Company could have been the Apollo pressure suit supplier from the beginning. While this was not to be, another opportunity presented itself.

In early 1962 the Mercury Mark II program was already underway having started the year before. NASA thought they had the ideal pressure suit approach with take-apart suits that had disconnects at various places to remove suit limbs to facilitate partially unsuiting in a confined cabin space to achieve long-duration comfort. The summer of 1962 saw two major changes in this program. First, on June 3 the program was renamed the Gemini program. Second, before NASA's suit approach could be fully developed, the David Clark Company internally funded a prototype suit and offered it to NASA for evaluation as an alternative approach to long-duration comfort. It featured a single zipper starting below the collar in the back, running down the back, between the legs, and up the front to slightly above the waist (Figure 4.1). The individuals responsible for this derivation of the Model 18 entry system were Richard "Dick" Sears and Donald "Don" Robbins.

Dick Sears joined the David Clark Company in the late 1950s to assist with the company's increasing pressure suit business. Sears quickly grasped the nuances of both partial and full-pressure suits, making valuable contributions to numerous suit programs.

Don Robbins was also a motivated and versatile designer who had a strong interest in the business. He was David Clark's son-in-law.

Their entry system provided the straights and gentle curves needed for the best structural performance of the zipper. This also made the suit easier to don and doff than the other NASA pressure suits of the time. The two-way zipper

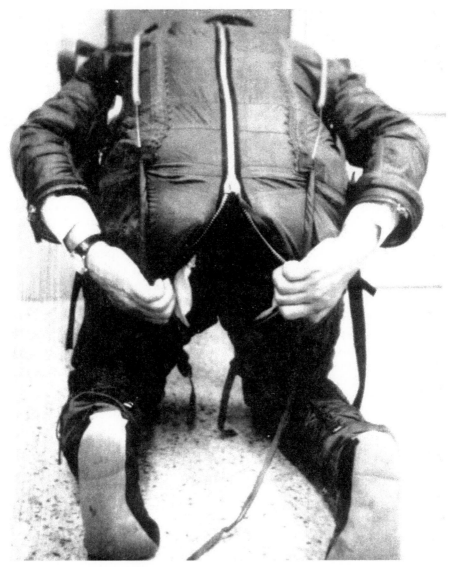

Figure 4.2. The rear-entry advantage
(courtesy NASA)

proved to have another advantage: by pulling down the zipper at the front the astronaut could easily address going to the "bathroom" (Figure 4.2). This was of critical importance in situations where an astronaut had to remain in his suit for a period of days.

The rear-entry David Clark suits were found to be superior to the rest regarding the Gemini requirements and were subsequently used for the Gemini program. David Clark production suits soon replaced the prototypes to support evaluations. They were used in early vehicle interface evaluations that were critical to spacecraft development and production. A second design, the David Clark Training Suit, soon followed; it incorporated lessons learned from earlier testing. The most noticeable feature of the first two designs of Gemini suits was the aluminized outer fabric that had been selected for their space vacuum thermal properties. During 1963 astronauts and other suit subjects were testing the first Gemini and Apollo pressure suits interchangeably. Many expressed a preference for the David Clark Gemini suit and its rear-entry system.

Overall problems with the Apollo design caused NASA to test Gemini suits against the full range of Apollo suits in early 1964 for Apollo use. Only the Gemini suits were found adequate for Command Module operations. At the conclusion of a Lunar Module evaluation at the Grumman facility on Long Island, New York, one astronaut stepped up to the podium to express his dissatisfaction with the Apollo suit and how he preferred his (David Clark) Gemini suit. While these were dark days for the Apollo Space Suit Program, these events led NASA to contract the David Clark Company to provide the suits for the early Apollo missions that did not include spacewalks. These missions were called Block I. NASA formally announced the decision in October 1964 with the Apollo Block I contract being formally signed in March 1965. However, March 1965 had much greater significance to space history.

Soviet Union Cosmonaut Aleksey Leonov's performed the world's first spacewalk on March 18, 1965, an event that abruptly changed the direction of the Gemini program. The first U.S. astronaut venturing out into space to perform an extravehicular activity or "EVA" had been planned for 1966. Richard Johnston, the Crew Systems Division chief (part of the then-Manned Spacecraft Center, now Johnson Space Center in Houston) convened a meeting nine days after Leonov's pioneering spacewalk. He told a small group of people that NASA wanted to perform the first U.S. EVA on Gemini IV, which was scheduled for early June 1965. The requirements were for a suit which could withstand the harsh space environment, a life support system, and a handheld maneuvering unit.

This required overgarments that could protect the astronaut from the blistering heat of direct sunlight (200°F, 93°C) and the freezing cold (−140°F, −96°C) of extended shade in space. Up to this point, NASA's Elton Tucker had provided the leadership by establishing the materials and ply orientations. However, the Gemini program had only progressed from test samples to a couple of prototype garments. The suit's single outer abrasion cover was replaced with layers of nylon felt for micrometeoroid protection, seven layers of aluminized Mylar for thermal insulation, and an outer layer of high-temperature HT-1 fabric. To make this possible the folks at David Clark, especially the master seamstresses on the fourth floor, had to complete development and deliver flight quality garments in less than eight weeks with no room for error. The techniques used by David Clark to manufacture insulation garments would later be shared with International Latex and Hamilton Standard in support of their Apollo efforts.

Figure 4.3. The first U.S. EVA performed by Ed White
(courtesy NASA)

The helmet had a two-visor system (both external). The outer visor was gold-coated to prevent harmful amounts of light and infrared energy that would otherwise blind the astronaut. The inner visor provided impact protection and thermal control. Thermal overgloves were also created to protect against conductive heat transfer.

Decades of craftsmanship in producing fabrics and the best talents of the David Clark personnel transformed the Gemini training suit design, which essentially had no insulation, into a spacewalking design that successfully supported America's first spacewalk. On June 3, 1965 Gemini IV Astronaut Ed White conducted his historic EVA or spacewalk (Figure 4.3) for 36 minutes outside the spacecraft. While there were difficulties, to the world at large this appeared to be a well-choreographed ballet.

The thermal overgarments and the pressure suit inside worked flawlessly. The EVA went so well that the U.S. spacesuit community was led to believe that zero-gravity spacewalking would be easy.

As the Gemini overgarments had proven themselves in space and the Apollo program was struggling with overgarment development, NASA prevailed upon David Clark himself to train first Hamilton Standard and later International

Figure 4.4. Early David Clark Apollo A1C suits
(courtesy NASA)

Latex personnel in spacesuit thermal overgarment construction. Thus, another element of Gemini suit development flowed into humankind's first footsteps on the Moon.

By the summer of 1965 the David Clark Company was delivering Apollo A1C suits to NASA. It not only seemed assured that the early manned Apollo flights would use David Clark suits but with the only two prototypes making the Apollo Block II competition deadline of June 15, 1965 being a Hamilton-Goodrich design and the David Clark completion prototype, many thought the David Clark Company suit would win making David Clark the suit supplier for all the Apollo missions. While this probably brought joy to the hearts of many who worked in Worcester, this was short-lived. NASA allowed ILC to enter the competition late and within a few weeks ILC turned out to be the winner. However, at this point the people of David Clark could still look forward to being a part of the first Apollo missions.

The A1C suits differed little from the Gemini suits (Figure 4.4). Unlike Gemini, the Apollo Block I helmet featured a fiberglass outer shell over the pressure and sun visors (Figure 4.4) for impact protection. The accidental sinking of the space capsule during the water landing of the second manned landing made astronaut flotation gear very important. Early A1C suits used a Hamilton-developed, chest-mounted flotation device (Figure 4.4).

David Clark Block I suit deliveries permitted Apollo vehicle development, crew training, and flight preparation to proceed in parallel with Gemini program operations. David Clark personnel compared the flotation perform-

Figure 4.5. Grissom, White, and Chaffee in late A1C suits
(courtesy NASA)

ance of early A1C flotation devices to that developed by the David Clark Company for Gemini. The comparative data were shared with NASA. By late 1966 the flotation device had been revised. This was now a David Clark harness system featuring flotation devices under each arm (Figure 4.5, shown stowed). Although the attachment method was revised, these flotation devices served all the Apollo missions by enhancing post-flight safety (Figure 4.6, shown inflated).

In 1965 the molded convolute joint of ILC effectively won the Apollo Block II lunar suit recompetition. In the course of the competition, someone from the David Clark company boasted that David Clark could make suits with molded convolutes. This sufficiently aroused NASA curiosity that NASA funded a David Clark prototype to see whether such a hybrid of David Clark and ILC technologies provided an improvement in pressure suit mobility. For this proto-type the David Clark Company manufactured mobility facsimiles of ILC Apollo convolutes and integrated them into a pressure garment. The resulting David Clark prototype (Figure 4.7) provided no overall improvement over the Gemini and ILC designs. However, NASA found the attributes of the system of interest and ordered a limited production of this design for technicians support-ing astronauts inside vacuum chambers. NASA gave this technician chamber suit the designation S-1C.

At the start of 1966 there were still some lingering hopes at the David Clark Company that they might win additional Apollo business. Based on the success of the Gemini IV spacewalk, many David Clark people expected the next and subsequent Gemini spacewalks scheduled for 1966 to be equally successful. If

Figure 4.6. Landing egress training in 1968
(courtesy NASA)

that occurred and ILC faltered in Apollo Block II, the staff at David Clark
might still have the opportunity to provide the pressure suits used by the first
humans to walk on the Moon. Unfortunately, the subsequent Gemini spacewalks proved just how difficult it was to work in zero gravity. Hence, any
hopes of providing the Moon suits ended.

On January 27, 1967 a fire during a capsule checkout claimed the lives of the
first Apollo crew Grissom, White, and Chaffee (Figure 4.5). The subsequent
investigation determined material changes were required throughout the vehicles
and spacesuits. Rather than funding the redesign of both the David Clark and
ILC suits, NASA judged the ILC suit ready to support the initial Apollo
missions (i.e., those without EVAs) and funded only ILC.

Not much in the way of redesign was going on at this point in the aviation
pressure suit business. The David Clark Company not only had experience in
spacesuit communications systems but also in defense applications and commercial aviation. This expertise led NASA to fund David Clark to develop a better
Communications Carrier Assembly (a.k.a. "Snoopy Cap") for the ILC Apollo
suit (Figure 4.8). Consequently, communication from astronauts on the Moon
would all be through a David Clark product. This design would later be used
by NASA for use in the suits used for Shuttle and International Space Station
spacewalks.

NASA went on to fund an advanced David Clark prototype (Figure 6.4),
but that represented the end of Apollo space pressure suit business. While the
management of David Clark tried to find alternative work for their "suit"
people, the lack of business forced the company to lay off most of the engineers
and seamstresses who had made the Gemini and Apollo Block I suits possible.

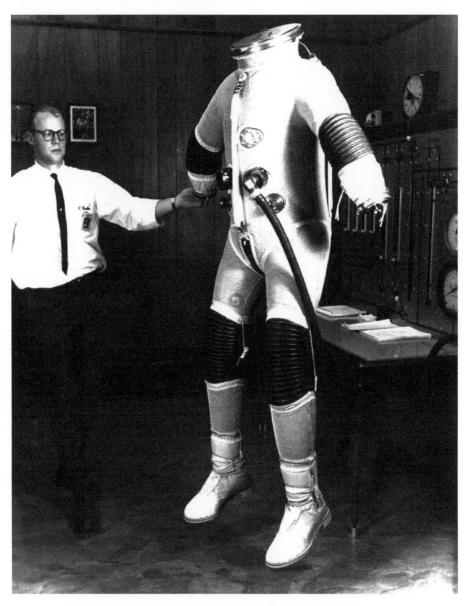

Figure 4.7. Jack Bassick with the Model S-1C Chamber Suit
(courtesy David Clark Company)

As ILC and Hamilton Standard were overflowing with work as a result of Apollo and Manned Orbiting Laboratory pressure suit contracts, respectively, both organizations elected to advertise in the Worcester area newspapers for experienced David Clark pressure suit workers. Given the help that the David Clark Company had provided Hamilton and ILC, this left long-lasting bitter

Figure 4.8. The Apollo 9 through 17 Snoopy Cap
(courtesy NASA)

feelings with many at David Clark. While some David Clark personnel migrated to Delaware and Connecticut to continue spacesuit employment, most chose to find alternative work in the area. When the David Clark aviation pressure suit business picked up, some returned.

The talent and experience of its Gemini/Apollo veterans coupled with younger associates allowed the David Clark Company to remain in the aviation suit business and reenter spacesuit work providing launch and reentry-type spacesuits for the Space Shuttle program. While those who worked in the Gemini and Apollo programs are now all retired, their contributions to David Clark Company practices, procedures, and technical knowledge continued and the David Clark Company remains a competitor in the spacesuit business of the future. However, this is not quite the end of the David Clark Company Apollo story as David Clark purchased Air-Lock in 1998.

Air-Lock was ILC's supplier of metallic suit parts in the Apollo program. The two companies made considerable contributions to the suits that were used to explore the Moon. The Air-Lock Apollo full-bubble polycarbonate helmet, which has seen only minor changes, still supports U.S. human space exploration today aboard the International Space Station (Figure 5.5).

CHAPTER 5

NASA's Block II Spacesuit Program and the Impact of Fire on Apollo

The results of the Block II competition left the Apollo Space Suit Program in chaos. The first three years of the program had demonstrated that ILC and Hamilton simply could not work together. Who was going to manufacture the Apollo pressure suits or be in charge of the overall program was unclear. Based on the quality issues ILC had had in the past, NASA's Apollo Pressure Suit Assembly Manager, Jerry Goodman, urged NASA to use the ILC design but have the suits manufactured by David Clark or Goodrich. This was not well received by Matt Radnofsky and Jim Correale. Goodman was soon replaced by James W. "Jim" McBarron II as the suit manager.

Jim McBarron hailed from Lima, Ohio. His interest in science led him to attend the University of Dayton. While there, he gained a part-time job during the school year at Wright Field as a test subject and technician supporting high-altitude protective equipment and pressure suits. This was a full-time job in summers and holidays and gave McBarron experience in X-15, X-20, and Mercury pressure suits. Not long after he graduated from the University in 1960, he was hired by Dick Johnston of NASA's Space Task Group at Langley Field, Virginia. McBarron followed Johnston to Houston to join Crew Systems. This settled that ILC was the Pressure Suit Assembly provider.

Following the Block II competition, NASA's first step as acting integrator was to conduct evaluations that were as realistic as possible of the complete spacesuit (Figure 5.1). NASA had created a simulated lunar surface on their Houston campus. This was not in an air-conditioned building but rather outside in the broiling August sun of Houston. To add to the misery the ILC competition suit did not have a liquid cooling garment. Jack Mays had to make frequent stops to rehydrate as the sweat pooled in his gloves and boots, which is illustrated by his pouring sweat from one of his gloves in Figure 5.2. The testing was conducted with an umbilical for pressure and air replenishment. This added an encumbrance despite numerous personnel moving the umbilical in unison with Mays' every move.

Figure 5.1. Assembling for evaluation, August 1965
(courtesy NASA)

The events of 1965 also produced a significantly different ILC. Gone were management decisions, oversight, and visible involvement from the corporate headquarters in New York. The new system integrators, NASA, dealt directly with Len Shepard and his Dover personnel.

By the start of the Apollo Block II Pressure Suit Assembly program, ILC-Dover's suit facilities had been completed. These were but a small part of the International Latex main Dover campus. To support Apollo, ILC had taken an existing building and stripped the interior to the bare shell, then designed an essentially new building in the existing shell. Every function was closely located to each other. Every aisleway, whether an engineering office or manufacturing area, was designed to allow forklifts and pallet jacks to easily bring in whatever was needed and take out finished products or process residue. The only suit-manufacturing function not part of this centralized area was the Dipping Facility, which was located a few blocks away from the campus that the program shared with other International Latex business units.

There was a short-lived NASA-ILC battle over boots. NASA liked the boots on the Hamilton-Goodrich competition suit which had a walking ankle joint. Moreover, NASA had had favorable experience with Litton suits with ankle

Figure 5.2. Jack Mays dumping sweat from his glove
(courtesy NASA)

joints. ILC argued their boot which had no ankle joint was better on the basis that it provided greater safety in that an ankle joint increased the likelihood of an ankle injury on the Moon. In this area ILC temporarily prevailed.

By the end of August 1965 NASA had informally assumed the role of Extravehicular Mobility Unit (a.k.a. "EMU" spacesuit) integrator and directed ILC to proceed with training suit manufacture in advance of a formal contract award. This left significant contractual issues including who was ultimately going to provide the backpack.

NASA could not formally issue contracts as the spacesuit integrator until Hamilton formally relinquished the contract. A further complication was that ILC wished to assume control of the Liquid Cooling Garment. Hamilton had internally funded the development and had a patent pending. NASA certainly wished to avoid all the potential disagreements that could arise from a Hamilton cooling garment being part of an ILC pressure suit. What is more, there was a financial element.

In Hamilton's 1962 proposal they had agreed to build dedicated Apollo test facilities through internal funding in Windsor Locks if they won the contract for the entire suit system. Hamilton were to be paid a nominal fee on all contract activities, both life support and pressure suit, for doing this. Hamilton did what they agreed and built the facilities. NASA removing the pressure suit and system integration roles from them meant lost business income. Since it was not realistic for Hamilton to gain compensation for what they volunteered to do in 1962, they were disinclined to surrender the Liquid Cooling Garment without considerable reimbursement.

Elements within NASA joined ILC in contesting the Hamilton cooling garment patent application on the premise that the RAF had invented the Apollo Liquid Cooling Garment. The ensuing battle and NASA's inability to issue formal contracts would not be resolved until March 1966.

The challenge to the Hamilton Liquid Cooling Garment patent was ultimately resolved by the Head of Crew Systems, Dick Johnston, hearing all sides and finding in favor of Hamilton's patent application. This set the stage for resolution of the Apollo contractual issues. In March 1966 NASA purchased the patent rights to the Liquid Cooling Garment from Hamilton allowing ILC to assume control of cooling garment manufacturing. Radnofsky was temporarily transferred to Washington and Hamilton formally surrendered the Apollo spacesuit prime contract. This enabled NASA to formally issue contracts to ILC for the Apollo Pressure Suit Assembly and Hamilton for the Portable Life Support System backpack. NASA made the rights to the Jennings cooling garment patent available to all U.S. corporations and individuals.

ILC's assuming responsibility for the Liquid Cooling Garments was not without challenges. While NASA purchased the patent rights, NASA elected not to purchase the Hamilton specification documents, which contained manufacturing and quality control details. Ronald J. "Ron" Bessette was tasked with recreating this information and ensuring ILC engineering documents reached NASA standards. This meant numerous trips from Dover, Delaware to the B. Welson & Company's facility in Hartford, Connecticut armed with drafts of documents so Welson's employees could help Bessette get the needed information. This had an upside for Bessette. He was from Moosup, Connecticut. He planned his trips so he would be returning on Mondays after spending the weekends visiting his folks. Bessette remained ILC's Liquid Cooling Garment Engineer throughout the Apollo program. This was another challenging period with NASA requiring additional development and quality control. Up until 1969 Welson continued to be the Apollo Liquid Cooling Garment manufacturer. In 1969 Welson elected to discontinue supplying cooling garments to ILC. Bessette had to teach ILC personnel so that they could assume subsequent Apollo production. Worldwide the spacesuit liquid cooling garments of today still bear a strong resemblance to those of the Apollo program.

In 1965 Hamilton internally funded the development of a second cryogenic ground test portable life support system that was named the "Liquid Air Pack," which reflected the use of cryogenic air rather than cryogenic oxygen. This was tested a short time later. Some 35 years later, Jack Mays vividly remembered how well the Liquid Air Pack worked (Figure 5.3). This enabled NASA to gas-cool test subjects and perform evaluations up to an hour and a half without an umbilical. This was vital to testing the spacesuit-to-vehicle interface needed to understand the state of the overall program. While evalua-

Figure 5.3. The Hamilton Liquid Air Pack supporting a Litton RX-2A test
(courtesy NASA)

tion of the Liquid Air Pack went well, NASA elected to make the subsequent production contract subject to competition, which AiResearch won.

The challenge facing the suit program now involved refinements and improvements to transform the design into a reliable, durable, and optimally performing garment. As part of the A5L Pressure Suit Assembly, NASA expected the provision of a helmet.

In 1965 and 1966 NASA's helmet effort led by Jim O'Kane and Bob Jones had explored a variety of vendors in the quest for optically acceptable helmet bubbles. They had not been successful forcing NASA/ILC to look elsewhere to support production. Air-Lock was contracted for the A5L Helmet which appeared to be a replica of an earlier Hamilton design. This proved to be a durable unit that supported Apollo suit production into 1967 and saw continued use in training and test applications (Figure 9.12) right up to the end of the 1960s and into the 1970s. Furthermore, Air-Lock produced prototype Lunar Extravehicular Visor Assemblies to support training and system evaluations.

By 1966 David M. Clark had sold his interest in the David Clark Company and was the proprietor of Air-Lock, having survived his founding partner Bill

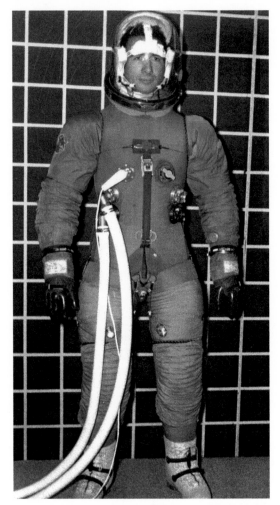

Figure 5.4. A Block II production suit
(courtesy NASA)

Boynton. Knowing of NASA's quest for an optical quality polycarbonate full-bubble helmet, Clark challenged his startup business, Clark Associates in Worcester, Massachusetts, which was operated by his son, Myron, to create such a product. To meet this challenge, Clark Associates engaged the services of Jerry Nault, who eventually met the challenge using his considerable skills. Upon making his first optical quality bubble, Nault took it to Clark to show it off, whereupon Clark remarked "Go make another one," which he did. Subsequently, Nault was hired by Air-Lock and the bubble helmet manufacturing process was relocated to Milford, Connecticut. This resulted in almost optical quality prototype Air-Lock helmets seeing use in the Apollo program in 1966 (Figure 5.4) and the production of optical quality helmet bubbles in 1967.

Figure 5.5. A U.S. International Space Station helmet
(courtesy NASA)

The David Clark Company handed the reins of Air-Lock over to Jim Edwards, Bill Boynton's son-in-law. Jim presided over Air-Lock's success in providing the high-quality spacesuit hardware for NASA's Gemini, Apollo, Skylab, Space Shuttle, and International Space Station programs (Figure 5.5). Along the way, Jim acquired Air-Lock from David Clark himself. In 1998 the David Clark Company purchased Air-Lock from Jim Edwards.

ILC also faced many challenges. In the Block II competition of 1965 ILC submitted the Torso Assembly which, coupled with a NASA helmet, completed the launch and reentry configuration of the pressure suit. ILC had to develop the extravehicular suit accessories that would permit astronauts to work in the thermal extremes of direct sunlight and shade in space. Another challenge involved the quality issues that had plagued ILC's suits under the Hamilton contract.

While the torso cover garments were being worked on by Robert C. "Bob" Wise, Richard C. "Dick" Pulling, and Iona Allen, the thermal overgloves posed such serious challenges that this development had to be supported by Durney's development group. In that effort one person stood out.

Rico was someone that people worked with but did not know well. One ILC employee believes her last name was Perry and that she came to the U.S. as a

Korean war bride. Most knew her only as Rico. What is well remembered is that she was an expert in origami. Rico could take a pile of blank paper and easily make a veritable zoo. This talent seemed to carry over into spacesuit materials. Rico was able to picture how to twist, turn, invert, and stitch the most complex fabric assemblies, which was a very unusual skill. This caused great challenges for the male engineers who followed her in the development process as they had to figure out how she did it and carefully take the assemblies apart to measure and document her creations so the production personnel could later fabricate the part. A great deal of Rico's skills was embodied in the thermal overgloves.

ILC conducted a quality improvement drive on their pressure suit, something that produced results. Gone were the rejections that had plagued earlier deliveries. The A5L suits were inspected and put into training without issues.

Anna Lee Pleasanton was the pattern maker, a key technologist striving for reliability and optimal performance. Her patterns guided the changes and provided subsequent process controls. Typically a drafting function, Anna brought understanding of how bias, an invisible orientation that is formed into the fabric in the weaving process, affects the form, fit, and function of every panel crafted into a suit.

The battle for refinements and improvements manifested itself in other areas as well. Mel Case was a dedicated engineer who was known for his honesty by everyone with whom he worked, both in and outside of ILC. He improved the molded convolute process by eliminating dipping, which not only improved manufacturability but decreased leakage failures and improved durability.

In 1966 it was decided that the life support backpacks would be left on the surface of the Moon to maximize the size of the lunar geological sample payload returning to Earth in the Lunar Module ascent stage. The possibility of the Lunar Module being unable to dock with the Service and Command Modules raised a new safety concern.

In 1966 the Apollo EVA suit system only had a 5-minute life support capability. A transfer from the ascent stage of the Lunar Module to the Command Module was estimated to take up to 30 minutes. This resulted in a 30-minute on-suit emergency life support system named the Oxygen Purge System (Figure 5.6), which was added to the top of the backpack (Figure 5.7). As part of its development, Hamilton convinced NASA to incorporate a regulator that offered two flow rates. The high flow rate delivered what NASA wanted, which was a 30-minute replacement for the main life support backpack. This not only supplied oxygen and removed carbon dioxide but also provided cooling from the expanding gas as it entered the suit's ventilation system. If cooling was not required, say in the case of a backpack oxygen system malfunction where cooling remained working, the Oxygen Purge System could supply 75 minutes of gaseous life support with some leeway.

In October 1966 David H. "Dave" Slack joined ILC as a reliability technician but rose in a matter of months to be a non-degreed reliability engineer, and completed his engineering degree during Apollo. He served in

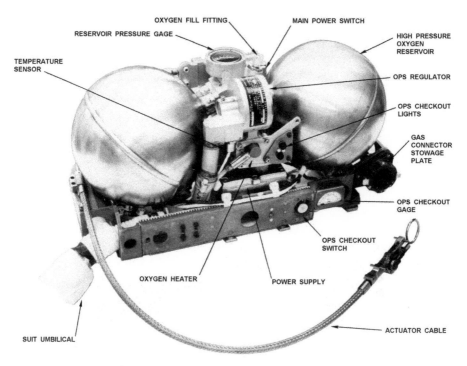

Figure 5.6. Oxygen Purge System components
(courtesy UTC Aerospace Systems)

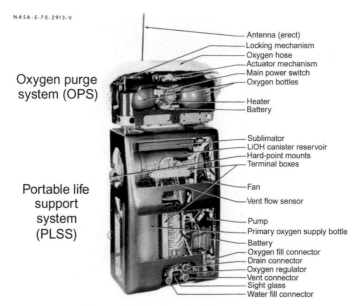

Figure 5.7. The OPS/PLSS relationship
(courtesy NASA)

that position until the summer of 1968. Slack was a tall and fit young man with a reserved demeanor rooted in impeccable honesty. He entered ILC's pressure suit program after completing high school and seven years in the Air Force. His position gave him insight into ILC developments and technical challenges.

In late 1966 the Apollo flights involving Lunar Modules and extravehicular activity were expected to start in less than a year. NASA prepared for pressure suit and complete suit system testing of the Apollo Block II configuration to validate that the suit system was ready for space and lunar surface mission use.

Although Apollo spacesuit testing led to many minor injuries, one test was nearly deadly: the test on December 14, 1966. A suited NASA test subject named Jim Le Blanc was inside a vacuum chamber at the Manned Spacecraft Center (now Johnson Space Center in Houston). The air was being drawn out of the chamber to simulate the vacuum of space. At a pressure equivalent to an altitude of 150,000 feet (46,000 meters) an oxygen hose became disconnected from the connector on the life support side of the spacesuit. As the connector was still attached to the pressure suit a backup safety valve within the suit's life support connector was unable to close because the external connector on the life support side was still in place creating a hole in the suit and exposing it to the almost vacuum inside the chamber. The suit quickly lost internal pressure. Le Blanc said that before he passed out he felt the saliva on his tongue start to bubble. Sealed in a vacuum chamber, Le Blanc would quickly perish. Clifford Hess, a NASA technician supporting the test, heroically repressurized the chamber in one minute. In doing so, he accomplished activities that normally took about half an hour. Jim Le Blanc regained consciousness almost immediately and was, thankfully, not injured. Hess and another NASA employee named Henry Rotter received certificates of commendation from Dr. Robert R. Gilruth, Director of the Manned Spacecraft Center.

The resulting investigation found the connector on the life support side was not a unit made by Air-Lock but one made in Houston to support testing. It was reverse-engineered incorrectly resulting in changes that caused the hose not to be securely attached to the connector. In less than two weeks the investigation was completed, the corrective actions were implemented, and man-testing was allowed to resume.

The next scheduled chamber testing was a complete suit system evaluation series for lunar environments. The tests were performed in the thermal vacuum space chamber at the Manned Spacecraft Center in January 1967. The chamber tests represented the most severe cold and hot conditions encountered during the Apollo missions. For these tests an in-chamber, overhead lifting system was planned so as to lift most of the spacesuit's 159-pound mass off the suit evaluator inside. This was to allow better simulation of lunar conditions. Regretfully, this system was not yet ready.

One of the test subjects was Richard A. Hermling. Hermling's path to NASA started with him joining the U.S. Air Force and graduating from physiological training school where he learned about pressure suits. He earned a certificate as an engineering technician before leaving the Air Force to join

NASA in 1966. Hermling was selected as a test subject because of his size and experience. Being six foot, large in size, and in excellent shape, he represented the larger astronauts in the program. Hermling was one of three subjects recruited for this complete system-level testing. The other two were William "Bill" Farley and Edward "Ed" Kuykendall. However, the number of text subjects soon reduced.

In a non-chamber pretest practice run in the spacesuit, Bill Farley was performing a mission simulation when he heard a popping sound in the suit. He immediately stood up straight and relayed over the communications system what he had heard. The suit engineer supporting the practice run replied that he thought it was just Velcro separating. Farley told the suit engineer "I know the sound of Velcro separating and that's not what I heard." Dissatisfied with the explanation and possibly the overall safety of the suit, Farley withdrew as a test subject after the run.

Hermling was to be the next suit subject but the medical staff would not approve him for the test as a result of a stomach ailment. Thus, Kuykendall became the suit subject for the first actual thermal vacuum test. However, as they were starting the test, there was a backpack cooling problem that caused the test to be aborted. The resulting investigation found that the cause was "tolerance stack-up." The manufacturing tolerance range on the cooling garment connector, the backpack connector, and the port on the front of the suit into which the connectors attached were all at or near the ends of their allowances in just the right directions to cause cooling water not to flow. Changing one of the connectors rectified the situation. Screening existing units and revising future manufacturing tolerances eliminated potential recurrence.

Before the next test, Hermling was approved for the first thermal vacuum qualification test after having a medical. This was a successful "hot" test of the spacesuit. Hermling reported "The paint was blistered on the helmet during this test. It was so hot; I could barely keep my hands in the gloves. Other than that, everything functioned as it was supposed to." Asked how it felt, Hermling described it as "like a hot, dry summer's day in Houston."

The next test came three days later with Ed Kuykendall as the test subject. This was to be the "cold" test. During testing a crotch cable broke on the suit. The pressure load was transferred to the fabric section of the leg which length-ened pulling Kuykendall's foot out of the boot but the garment retained life-supporting pressure. Kuykendall had to make his way off the test setup and out of the chamber with a series of hops. This failure caused Kuykendall suffi-cient safety concerns that he too withdrew as a test subject. Testing was delayed a few days to allow repair of the pressure suit. The subsequent corrective action was a refinement of the cable-manufacturing end-connector crimping process.

The final test, a repeat of the cold test, came four days later on January 27, 1967 (Figure 5.8). Richard Hermling was the test subject. The test started with all appearing to perform well and only minor problems through most of the activities. However, before the end the lithium hydroxide carbon dioxide removal canister stopped functioning and significant fogging occurred in the

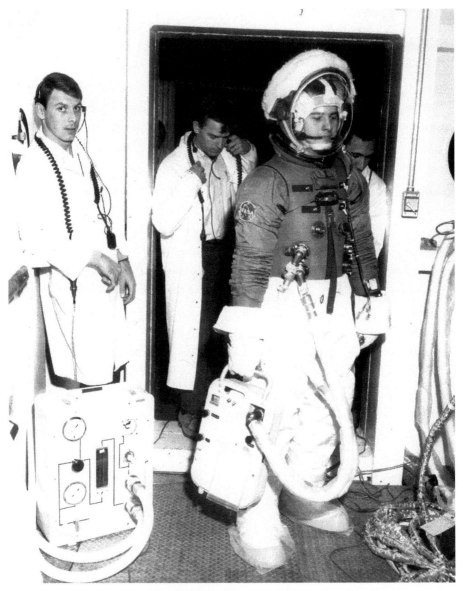

Figure 5.8. The "Cold" test
(courtesy Richard Hermling/NASA)

helmet. The test was stopped. This failure was subsequently traced to improper field processing of the carbon dioxide removal cartridges, which was consequently rectified.

In the 1960s astronauts (Figure 1.7) were American heroes. Most were former military test pilots who boldly pushed the limits of experimental aircraft permitting humans to go faster and higher in the technology drive leading to the space race. Later on the same day (January 27) Astronauts Grissom, Chaffee, and White (Figure 4.5) were killed in a flash fire in their spacecraft cabin during a test on the launch pad. This tragedy rocked the space program and the nation. The American Manned Space Program initiated a "stand-down" where everything stopped for 21 months while NASA and all Apollo contractors reviewed all the problems experienced, identified the root causes of the anomalies, and implemented robust corrective actions to the extent possible for any potentially major safety issue. Additionally, all possible failure causes of all components and systems were reviewed. Their potential impacts were identified and, where possible, the opportunity for failure reduced or eliminated. The working hours went from long to all consuming in Houston, Dover, and Windsor Locks. Despite manufacturing coming to an abrupt stop, engineers almost lived onsite until the late summer of 1967. Then, engineering settled into hard days overflowing with deadlines but at least having reasonably predictable working hours. By contrast, production at Dover and Windsor Locks went from finding alternative work so as to retain skilled people to getting everyone back, authorizing overtime, and working around the clock. That would go on for yet another year.

Corrective actions brought many material changes that required recertification. It was in this area that Matt Radnofsky, who had returned from his sabbatical in Washington headquarters, proved to be an invaluable aid in the selection of replacement materials. As NASA deemed the ILC design capable of supporting all phases of the Apollo mission, NASA chose to cancel the Block I pressure suits manufactured by the David Clark Company to concentrate material and process improvements on the ILC A6L suit. Temporarily, NASA's Block III advanced suit effort was allowed to continue in parallel.

Beyond corrective actions the Apollo Fire Quality and Safety Review provided the opportunity to refine every facet of manufacturing, field handling, and flight preparation. This resulted in the suit and backpack reliability that typified the flight portion of the Apollo Space Suit Program.

CHAPTER 6

The Cancelled NASA Apollo Block III Advanced Suit Program

The grand finale to the Apollo program and the crown jewel of mission scenarios was Apollo's Block III with each mission being a month and a half in space and every visit lasting over one month on the surface of the Moon. The need for a spacesuit that could support such an endeavor extended the life of Apollo's advanced suit effort.

The Houston-based advanced spacesuit effort started as a "contingency" to Hamilton Standard and International Latex Corporation (more commonly called ILC) spacesuit development of the main Apollo Suit Assembly Program. In 1963 that program was in trouble. Joe Kosmo was the NASA engineer selected to lead the advanced suit effort. The contingency part of the advanced program was a ruse. The initial Apollo vehicles being developed were incapable of supporting the type of suits Kosmo would develop. However, the ruse provided the funding. The real purpose was twofold. First, any parallel effort placed added pressure on Hamilton and ILC to succeed. Second, everyone knew that setting foot on the Moon was but a first step. More extensive human exploration would follow and for that more advanced spacesuits would be needed. So, Kosmo (Figure 6.1) was funded to play in an engineer's dream sandbox.

In October 1964 NASA top management planned to order 15 flightworthy Apollo Saturn V rockets, and the Apollo suit programs became realigned to Apollo vehicle systems. What had started out as the Apollo Space Suit Assembly program became Apollo Blocks I and II. Block I would use slightly modified Gemini suits, as they were proven to be able to support that Apollo phase. Block II would use whatever suit won a competition in July 1965. The Apollo Block III vehicle configuration was to provide a 35-day lunar visit. There would be almost daily routine spacewalks as opposed to the limited two to four surface outings of the earlier Apollo missions. For that NASA needed the best spacesuit possible. So, advanced suit development continued. While the intricacies of the funding of Block III had not been communicated to Kosmo, this provided continued funding for his efforts.

Figure 6.1. A young Joe Kosmo in a Litton RX-3 suit
(courtesy Gary L. Harris)

The Apollo Block III spacesuit story was not just about NASA. It was a tale of space community competition that was dominated by two California companies, Litton and AiResearch.

The funding for the design and development of the Lunar Module to support such long durations flowed in parallel with the spacesuit effort. However, the Block III vision did not manifest itself in the funding of a Block III program office to identify and coordinate all the related activities to bring this plan to fruition. Instead, NASA elected to fund a few activities with long development times, such as advanced spacesuit developments, with the intent of providing the funding for a program office to guide the overall effort and perform the remainder of the program activities at an appropriate time in the future. Joe Kosmo would not even hear the term "Block III" in relation to his suits until a few years later.

During 1965 to 1967 Kosmo directed Litton in a series of design iterations such as the RX-2A, RX-3, RX-4, RX-5, and RX-5A. Each design iteration

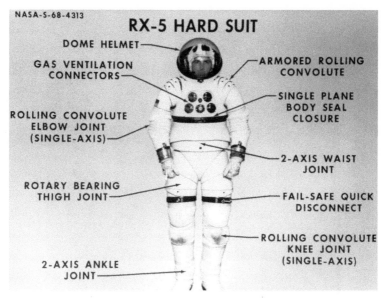

Figure 6.2. Litton RX-5 features
(courtesy NASA)

demonstrated some key improvement. The RX-5A (Figure 6.2) had mobility that far surpassed the capabilities of the other pressure suit manufacturers. This was the result of teamwork between NASA and Litton engineers and technicians all working together and contributing to the overall cause. It was also a testament to the abilities and dedication of Litton personnel. Engineers such as Pierre Brusso, Don Taylor, and Bill Elkins distinguished themselves while technicians Buck Scott and Jerry James made significant contributions.

However, Houston Crew Systems' advanced suit effort was not NASA's only Apollo era advanced suit program. When the initial Apollo spacesuit contract suffered setbacks, the Ames Research Center elected to join the fray. The Ames Research Center had previously been the Ames Aeronautical Laboratory and was established in 1939. Named for the chairman of the National Advisory Committee for Aeronautics, Joseph S. Ames, this center was located at Moffett Field in Sunnyvale, California. The Laboratory was renamed with the formation of NASA in 1958. Ames became a NASA spacesuit design and fabrication center by a coincidence coupled with a mid-management decision, rather than by planning at NASA headquarters. This story starts with a young engineer named Hubert C. "Vic" Vykukal (Figure 6.3).

Vykukal, who graduated from Texas A&M in 1959, went to work in industry for three months, and then entered the U.S. Air Force to fulfill his military obligation after graduation. His first assignment was to the Air Force Flight Test Center's White Sands Proving Grounds in Alamagordo, New Mexico. He was assigned to the High Altitude Test Chamber Branch. Five months after he arrived, he was reassigned by the Air Force to NASA Ames

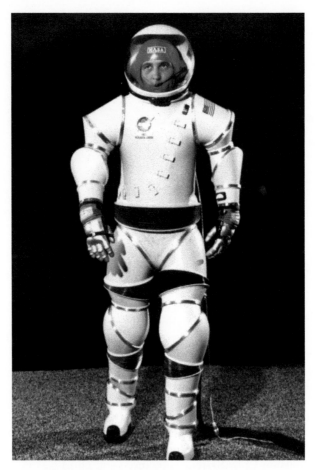

Figure 6.3. Vic Vykukal in the Ames AX-1
(courtesy NASA)

Research Center where he joined the Flight and Systems Simulation Branch to participate in pilot g-tolerance studies. Upon completion of his tour of duty, Vykukal was hired by Ames to work in the newly created Life Sciences Directorate.

In 1964 Vykukal's boss, Steven Belsley, attended a conference in Houston where he watched a demonstration of a spacesuit. He was unimpressed. Upon his return, he called Vykukal into his office and said "go design and develop a space suit using materials other than fabrics. A Stearman biplane [a fabric-covered airplane] can fly but it can't go supersonic." And with that, Ames joined the small community developing spacesuits.

In contrast to pressure suits made of fabric, Vykukal's efforts produced prototypes with no fabric. Instead, he used innovative new technical approaches such as all-metal bellows. Unlike the government-contracting approach of the

Manned Spacecraft Center programs, the Ames prototypes were accomplished through in-house fabrication with only the metal bellows and off-the-shelf bearings being supplied by outside vendors.

Although not readily apparent, the decisions being made on the Lunar Module program were driving the future of Block III spacesuit implementation. Apollo Block I missions did not use Lunar Modules. By the end of 1967 the first 12 Lunar Modules had been ordered with three-day life support systems. The last three were ordered with provisions for rover vehicles but not long-duration life support systems or storage of advanced pressure suits. Thus, NASA had decided between 1964 and 1967 to have two phases of human exploration by Block II-type suits. The Lunar Rover Vehicle was to start service with Apollo 15 without more advanced pressure suits. While this appears to have not been communicated to any great extent, the Lunar Modules planned for Apollo missions 18, 19, and 20 also did not support advanced Block III suits and 35-day Lunar Module surface explorations.

The RX-5A ended the RX series. The RX prototypes were called "hard suits" because the main sections of the suit were made of metal or composites. Joints containing fabric allowed suit mobility. This meant that, when not in use, the suits functionally took up the space of a person. Two suits being transported to the Moon equaled the space of two extra people in a vehicle. The Command Modules on order could not support the additional space needed.

The volume impacts of a hard-suit approach were recognized in 1967. NASA performed a survey of all the available spacesuit technology concepts. This included the upper torso of the Ames AX-1 (Figure 6.3)—the lower torso had not been completed. The AX-1 concept was not aligned with the new goals of Block III since the all-metal AX-1 was the ultimate in a non-collapsible suit. NASA also funded the David Clark Company for an advanced concept suit that was completed in November 1967. This "soft-suit" prototype was officially David Clark Company Model S1021 (Figure 6.4).

NASA testing indicated that Litton was ahead of the field in mobility development, which resulted in Litton being directed to produce a soft suit with advanced mobility features. The soft suit meant that the non-flex portions of the suit would be fabric to allow the suit to be collapsed to a smaller size for flight storage. The resulting prototype was the Litton Constant Volume Suit (Figure 6.5), which was delivered in 1968. While the suit demonstrated exceptional mobility, the locations and orientations of the metallic hard details detracted from the ability to compress the suit volume for storage. Thus, a further design iteration was authorized. This opened the door for competitors.

In 1964–65 the AiResearch Division of Garrett Corporation (now Honeywell) was funded for spacesuit life support development parallel with the Hamilton backpack. This was to provide real program contingency, as NASA did not care who provided the backpack just so long as a working system was ready for the first lunar surface missions. With Hamilton delivering a backpack that met all the requirements in late 1965, funding of the AiResearch backpack

Figure 6.4. The David Clark S1021 suit
(courtesy National Air and Space Museum SI)

was terminated. The AiResearch people in Torrence, California desperately wanted another opportunity to get back into spacesuits.

In 1967 AiResearch elected to expand its capabilities to include entire spacesuit systems by developing an in-house capability for space pressure suit design and manufacture. To do so, AiResearch successfully recruited Bill Elkins and Jerry James from Litton and drew from its internal talent base for additional resources. To demonstrate its ability to design and manufacture a state-of-the-art space pressure suit, AiResearch independently funded the EX-1A prototype (Figures 6.6 and 6.7). To avoid potential litigation for

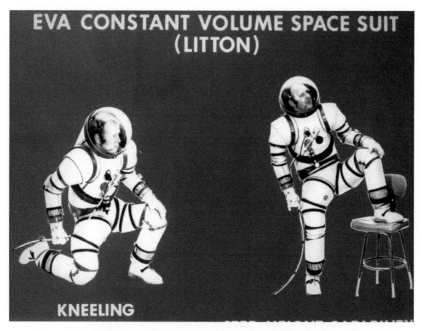

Figure 6.5. The Litton Constant Volume Suit
(courtesy Gary L. Harris)

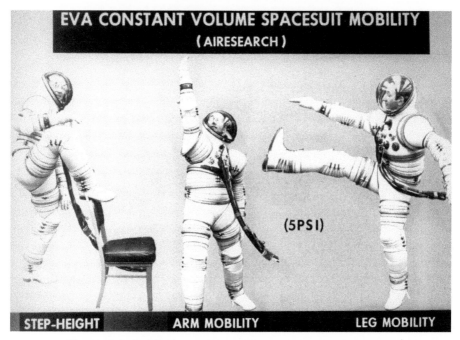

Figure 6.6. Bill Elkins demonstrating the AiResearch EX-1 suit
(courtesy NASA)

Figure 6.7. The EX-1 with thermal covers installed
(courtesy Gary L. Harris)

proprietary infringement AiResearch had to develop an all-new mobility system under tremendous time constraints. Their success led to the EX-1 being evaluated against Litton's Advanced Extravehicular Suit or "AES" prototype (Figures 6.8 and 6.9) in 1968. Comparative tests showed that AiResearch could compete with Litton.

The mobility performance of these two advanced suits was incredibly similar. The key distinction that won this face-off for AiResearch was the compact storage ability of its prototype (Figure 6.10).

By 1969 the space program was no longer the national priority that it had been in the early 1960s. The costs of the war in Viet Nam coupled with the domestic agenda of the Johnson administration's "Great Society" had created financial burdens that resulted in reductions of NASA budgets. Adjustments in NASA's strategic planning resulted. Kosmo was not allowed to fund any further suit developments. However, Litton did not leave the advanced pressure suit field without a fight, and while Litton's pioneering efforts failed to garner it any program production contracts, it left an influential legacy on what followed.

Litton was also developing an all-fabric "flat-pattern" joint system in 1969

Figure 6.8. The Litton AES without covers
(courtesy Gary L. Harris)

which Joe Kosmo found most interesting. This joint appeared to offer relatively low operating effort with a simple, less expensive manufacturing technique compared to the advanced mobility elements or molded suit joints of the time. Unfortunately, Kosmo could not get funding to explore this development and keep Litton in the suit development game. Litton disbanded its spacesuit group. Just a few years later, efforts resumed into NASA in-house and NASA-funded external suit mobility developments. However, this was with the help of

Figure 6.9. The AES ready for spacewalking
(courtesy Gary L. Harris)

both David Clark and ILC. The resulting design subsequently influenced the designs of the Shuttle and International Space Station Extravehicular Mobility Units.

The last known Litton pressure suit currently resides in the National Air and Space Museum's preservation collection. While mobility systems were common to the AES, the Litton suit featured a fiberglass hard upper torso and mid-entry

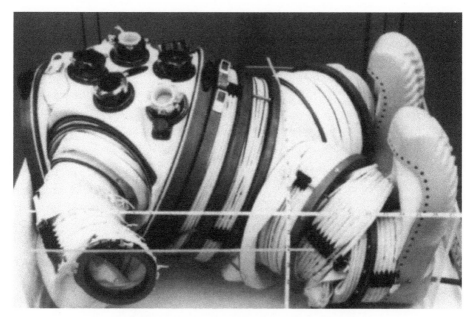

Figure 6.10. The AiResearch suit compressed
(courtesy Gary L. Harris)

system that were amazingly similar to what would eventually be used in the Shuttle spacewalking suit that started service in 1981 and continues to support NASA today.

Apollo Block III competition between Litton and AiResearch was not just restricted to pressure suits. NASA also funded limited life support activities.

In 1967 NASA funded Litton for the development of the Open Loop Portable Life Support System. This system (Figure 6.11) was designed around a "breathing vest" that was capable of capturing the chest volume in such a way that the expansion and contraction of the chest and abdomen would drive a bellows to move ventilation gases through the helmet. Between venting life support gas to space and the breathing vest, the resulting life support system would not need a fan, battery, carbon dioxide removal cartridge, or humidity removal subsystem.

AiResearch elected to compete against Litton for the advanced Apollo primary life support backpack and won the competition. The AiResearch back-pack was named the Optimized Portable Life Support System (Figure 6.12). This was an advanced self-contained system designed along the lines of the Block II backpack but featured a means of extending the time an astronaut could be supported during an Earth orbital missions or extended lunar surface extravehicular activity. This AiResearch life support system had on-mission checkout and servicing plus a cooling system that was regenerable, thus did not require boiling water to space. These features allowed a greater number of

Figure 6.11. Joe Kosmo testing the Litton backpack
(courtesy NASA)

spacewalks per mission than the Block II backpack. In addition, NASA envisioned funding AiResearch to develop a family of incrementally improved backpack life support systems.

At NASA's request, Litton also started development of a 45-minute version of the Open Loop Portable Life Support System, called the Contingency Transfer System for use during emergency return from the Lunar Module to the Command Module during proposed post-Apollo 17 missions.

On January 4, 1970 NASA announced the cancellation of Apollo 20 so that it could use the launch rocket to place the Skylab space station into orbit. Then in April 1970 Apollo 13 had an inflight failure causing the lunar landing to be aborted. The Fra Mauro landing site was reassigned to Apollo 14 and other landing sites were rearranged. On September 2, 1970 NASA announced it was cancelling the two remaining Block III flights as a result of further NASA budget cuts.

Neither of the Block III life support systems completed certification. The AiResearch Apollo 18 and 19 pressure suit was to be flown and tested as a flight experiment on Skylab. Unfortunately, complications with the Skylab launch and subsequent repair missions coupled with Skylab's fiery reentry

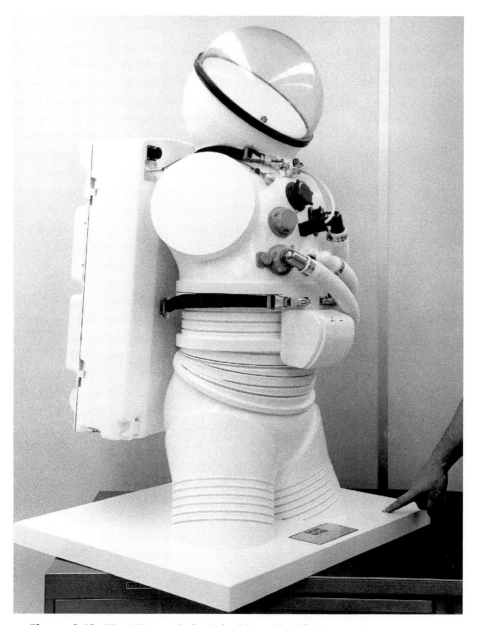

Figure 6.12. The AiResearch Optimized Portable Life Support System prototype
(courtesy Honeywell Corporation)

before the Space Shuttle could become operational resulted in those plans never reaching fulfillment.

Although design models of the Apollo Advanced Space Suit never reached flight, design features from the Advanced Apollo program did see space service in the Shuttle Extravehicular Mobility Unit or "EMU." Specifically, Advanced Apollo lineage could be found in the hard-ring torso closure, the "flat patterned" mobility elements, the modular suit elements with adjustable sizing, and integration of the backpack life support system with the hard upper torso. Advanced Apollo efforts also continue to provide design reference points in the 21st century.

CHAPTER 7

Final Spacesuit Efforts for Going to the Moon

The Apollo capsule fire safety stand-down culminated in a race against time. NASA wished to be flying Apollo missions in months. The selection of materials was just the first step. Then there was getting the materials or components containing the selected materials delivered, manufacturing the first articles, proving through certification testing that the new items met all the requirements, and delivery of suits and backpacks to support final training and flight. This put the certification process on a flight-by-flight basis. While the process had to be methodical and controlled, it was understood that there was no room for error and every day counted. There were thousands of items that had to be tracked to culminate in the first Apollo spacewalk. ILC probably had the greatest challenge as they had to produce a record number of suits in a limited number of months. NASA had to ensure that all the mission ingredients came together and everyone was thoroughly trained. Hamilton had produced a record number of spacesuit backpacks by remanufacturing the limited number of existing units and creating additional new production.

ILC'S APOLLO PRESSURE SUIT DEVELOPMENTS AND PRODUCTION

NASA needed a fleet of new suits quickly, but Apollo pressure suit manufacture was extremely labor intensive. Production involved intricate processes. Every step of manufacture required demanding attention to detail. The suits had over 20 layers. Then, there were space-level quality controls and documentation requirements. All in all, it took roughly 1,000 man-hours of effort to create each suit. ILC's workforce successfully responded to that challenge and, before Apollo 14 launched, at least 105 Apollo 7 through 14-type suits were delivered. This was an amazing feat. There was also an economic bonus, which took various forms, for those at ILC involved in Apollo suit creation.

The tidal wave of work offered almost unlimited overtime to key labor, especially hourly employees. However, it was not a tidal wave for all. This was the beginning of a support phase and a return to "regular working hours" for salaried engineers. Now these men had free time in the evenings and on weekends. Many elected to engage in entrepreneurial activities. Two engineers ran a nightclub in Dover in their off hours.

The safety stand-down not only gave the program the time to improve safety and reliability, it also allowed NASA to gain improvements to the Pressure Suit Assembly that time constraints previously did not allow. Specifically, the suit got a boot with an ankle joint and the cover garment became laced on or "integrated" with the torso assembly throughout the mission.

The person who made the Apollo boot a success was A. J. "Al" Kenneway. He not only had to add an ankle joint, he also had to make the boot work and feel like it was not under pressure. Shoes and boots needed to fold between the toes and the instep to allow for a longer walking stride. Pressure inside the boot wanted to make this area ridged and naturally caused the bottom or sole of the boot to balloon out. So, the sole needed to flex easily in one direction but be ridged and flat in the other. To some extent, Kenneway's efforts were a family effort as his wife Ceil worked in the sewing room.

The pressure suits used on the Apollo missions had Integrated Thermal Meteoroid Garments or "ITMGs." Prior to the Apollo capsule fire the outer garments were donned separately. As the name implies, the cover garments were integrated (i.e., attached to the pressure suit) throughout the mission, which resulted in greater wear on the fabric assemblies. Throughout both the Hamilton and NASA Apollo contracts, ILC Apollo suits were manufactured at ILC's Pear Street facility in Dover, Delaware. To support the great push to create the suits needed to support Apollo training and flight, more floor space was needed. As a result a factory in Frederica, Delaware was leased in 1968. The Frederica plant manufactured the Apollo ITMG outer garments through-out the contract while the Dover facility made the remainder of the Pressure Suit Assembly.

There were two different configurations of the Apollo 7 through 14 suits. The suit used by the Lunar Module crew had material and feature differences specific to surface exploration, including a liquid cooling garment. The Command Module Pilot used a suit with somewhat different materials and a Constant Wear (comfort) Garment in place of a cooling garment.

The Apollo pressure suit (Figure 7.1) was made from a variety of materials that were essentially constructed in layers. Each layer was selected or designed to perform specific functions.

ITMG materials that had the lightweight and thermal insulation properties needed to make lunar exploration a success were also fragile. Problems stem-ming from an outer garment getting damaged during a Gemini spacewalk demonstrated that as much durability and reliability as possible had to be crafted into this garment. The people whose efforts made this possible were Bob Wise, Dick Pulling, and Iona Allen.

Wise was a "Buckeye," a nickname for people from Ohio. His industrial Midwest attitude brought leadership and perseverance to the effort. This was supplemented by Pulling's dedication and support. Neither man would have

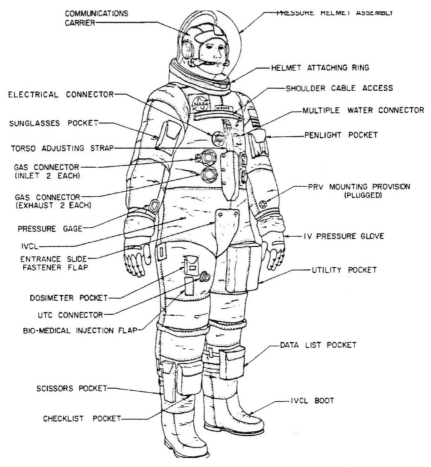

COMMUNICATIONS CARRIER

PRESSURE HELMET ASSEMBLY

ELECTRICAL CONNECTOR

SUNGLASSES POCKET

TORSO ADJUSTING STRAP

GAS CONNECTOR (INLET 2 EACH)

GAS CONNECTOR (EXHAUST 2 EACH)

PRESSURE GAGE

IVCL

ENTRANCE SLIDE FASTENER FLAP

DOSIMETER POCKET

UTC CONNECTOR

BIO-MEDICAL INJECTION FLAP

SCISSORS POCKET

CHECKLIST POCKET

HELMET ATTACHING RING

SHOULDER CABLE ACCESS

MULTIPLE WATER CONNECTOR

PENLIGHT POCKET

PRV MOUNTING PROVISION (PLUGGED)

IV PRESSURE GLOVE

UTILITY POCKET

DATA LIST POCKET

IVCL BOOT

Figure 7.1. The Apollo 9 through 14 pressure suit
(courtesy NASA)

achieved the success of the lunar ITMG were it not for Allen. She was a petite, slender African American woman who was a master seamstress. Her knowledge and craftsmanship aided Wise and Pulling's efforts through every phase of development and production. Together, their crafting magic made Apollo ITMG possible.

The outer layer of the ITMG was made up of woven glass fibers called Beta Fabric. This would not burn. It would only melt at temperatures substantially greater than those at which most of the materials in the capsule would burn.

The fact that Beta Fabric (used as the outer fabric material) was made of glass fiber seriously limited its abrasion resistance. ILC introduced the more durable Super Beta Fabric, which was a Teflon-coated Beta Fabric, on all Apollo suits starting with Apollo 10 to improve useful life. However, the life span of the suit's outer fabric was still limited. The nested insulation layers of

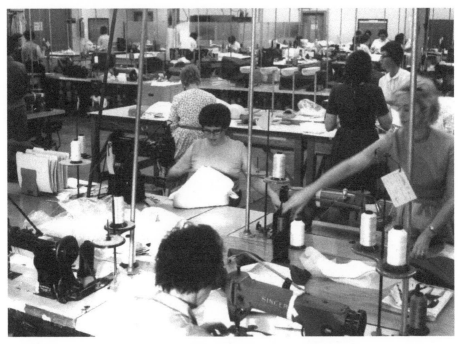

Figure 7.2. ILC Ladies at work for Apollo
(courtesy ILC-Dover LP)*

the cover garment proved to be extremely labor intensive, thus expensive, to produce. To enhance the flame resistance of the suit's cover garment the outer two layers of Mylar were replaced by two layers of Kapton sandwiched between layers of Beta Fabric marquisette. The Beta Fabric marquisette was bonded to the Kapton to retard breakage of the Beta Fabric insulation.

The addition of numerous pockets allowed the outer-garments of the Pressure Suit Assembly to be worn as a kind of "toolbox" to carry all the items astronauts needed for lunar exploration. Each suit involved hundreds of hours of work and was painstakingly crafted by highly skilled workers (Figure 7.2). Every astronaut had three suits made as identically as possible. Every section was cut out and sewn within a sixteenth of an inch. The threads per inch were carefully controlled on every seam to assure optimum structural integrity. To assure the seams would be in the exact locations in the sections, the latter were pinned together for pre-sewing alignment. As a pin left in the garment could cause the garment to fail in use, each seamstress was issued pins of a specific color. The pins were supposed to be accounted for at the end of each day. One story handed down among the engineers involved a master seamstress who found a pin in a garment while checking production. The color of the pin revealed the appropriate seamstress, who was summoned. When shown the garment and pin the seamstress accepted responsibility. The master seamstress then stuck the pin in the seamstress and ordered her back to work.

Figure 7.3. The crew of Apollo 7
(courtesy NASA)

The first manned flight was Apollo 7. The prime crew for Apollo 7 consisted of Astronauts Donn F. Eisele (Command Module Pilot), Walter M. Schirra Jr. (Commander), and Walter Cunningham (Lunar Module Pilot) (Figure 7.3, left to right respectively). It took a few months from NASA finalizing and approving the revised material selections to the delivery of the new suits needed for Apollo 7 preflight training.

Suits had to be constructed and certified in these months in parallel with custom fabrication of the Apollo 7 pressure suit sets. Each Apollo 7 set contained three identical suits for each member of the prime (Figure 7.3) and backup crews. Serial numbers clarified the relationship between the flight, training, and backup suits for Apollo 7. It was a matter of luck which suit was chosen as the astronaut's flight, training, and backup suits for subsequent Apollo flights. The first suit a Houston technician unpacked and put on the lab table for each astronaut became his flight suit. The second suit to be unpacked became the training suit. The third suit of the set became the backup suit. Once the serial numbers were recorded and the "training suit" label marked as such, the flight and backup suits were taken to a locked storage area until they were needed.

The personnel at ILC met this and subsequent flight challenges, ultimately producing at least 105 Apollo pressure suits for the Apollo 7 through 14 non-Rover missions. Given the design and NASA quality requirements, each pressure suit represented over 1,000 hours of human effort before delivery. In this race against time the production sewing area was deemed the highest skilled (Figure 7.2). In this area, three members of the ILC team stood out: Eleanor "Ellie" Foraker, Evelyn "Ev" Kibbler, and Roberta "Bert" Pilkenton. Ev and Bert were sisters. Foraker was chosen as the Seamstress Group Leader because of her tremendous sewing skills coupled with her have-to-get-it-done attitude. She was not only key to Apollo suit production but also later to saving the Skylab space station. Foraker was a widow and single parent who juggled the challenges of work and home. Kibbler was a like-minded and skilled sewer; a go-to person when you had a problem. There was no surprise when Foraker was promoted and Kibbler became the Seamstress Group Leader after Apollo. Pilkenton was a production area main sewer. Her abilities and work ethic made her stand out. The leadership provided by these three women in the sewing area made deliveries of the Apollo 7 through 14-type suit possible. As Pilkenton was Kibbler's sister these lunar suits represented a family endeavor.

One interesting element of these women-leaders is that they brought an element of extended family to this Apollo program. The experienced seams-tresses mentored not only the new sewers, cutters and pattern makers as the program grew to support NASA's needs but also new (in Apollo, male) engin-eers who in many cases rose into management. Reflecting the culture of time, this consideration was not always reciprocated. Later in the program, Hamilton Standard processed a series of patents to gain recognition from NASA for internally funded developments for U.S. space programs. This appears to have irked someone in ILC management. ILC drafted a patent application that seems to have been crafted to annoy Hamilton in return. The ILC portrayed itself as the inventor of the Apollo spacesuit crediting all Hamilton develop-ments to competitors or other organizations. It is interesting that the managers, in creating a list of ILC inventors, exclusively listed male employees.

The training both on land (Figure 7.4) and at sea (Figure 7.5) was intense. By 1968 mission simulations could be performed in air-conditioned buildings (Figure 7.4). Every aspect of the mission was repeatedly rehearsed. The impor-tance of post-landing activities (Figure 7.5) was well understood by everyone on the program as a result of the space capsule sinking on the second Mercury flight. Gus Grissom's pressure suit filled with water. Gus nearly drowned before the recovery rescue team reached him.

Delivery of the Apollo 8 suits brought to light an interesting helmet problem. Despite Frank Borman's head being measured for NASA's development of the Apollo helmet, the helmets did not fit him well enough. This meant Air-Lock had to develop and certify custom helmets and visor assemblies to support the Apollo 8 Commander. This further required the creation of all-new tooling. Given when the error was discovered, this put the mission instantly behind schedule. Air-Lock agreed to an almost impossible recovery schedule. Fortu-nately for the mission Air-Lock's resourceful people in Milford, Connecticut overcame the challenge so well that mission activities were not impacted.

The Apollo training suits were used well beyond their designed useful life. As there was a strong desire to keep the backup suit as a spare flight suit, the

Figure 7.4. Training in Houston's Building 19
(courtesy NASA)

Figure 7.5. Apollo post-landing training
(courtesy NASA)

training suits were frequently used and repaired. By flight time most of the training suits "had been used to death," and after the flight they were cannibalized for metal parts to support the construction of special one-of-a-kind evaluator suits, lab fixtures, or used for repairs on other training suits.

Apollo 7 and 8 in October and December 1968, respectively, used Apollo Pressure Suit Assemblies as launch and reentry emergency suits. In parallel with preparation for these two missions, certification demonstrated this suit configuration could additionally support requirements on the lunar surface. Apollo 7 Astronauts Schirra, Eisele, and Cunningham circled the Earth 163 times performing prove-out exercises of the Command and Service Module before returning. On the second manned flight, Apollo 8, Astronauts Borman, Lovell, and Anders circled the Moon.

However, Apollo 7 and 8 represented just the beginning. They made use of little more than a fraction of the Apollo suit system. Apollo 9 introduced the need for ancillary "spacewalking accessories" from ILC and life support backpacks from Hamilton Standard. There was also another significant development on the suit side of the Apollo program before humankind first stepped on the Moon: ILC no longer stood for International Latex Corporation.

In January 1969 International Latex chose to spin off its pressure suit and government products division by partial sale. The new and separate entity became ILC Industries. Many of the people who had supported the Apollo suit program were forced to decide whether they wanted to continue their careers with ILC or work on less exciting products for International Latex. Most chose to continue with ILC. The management of ILC suit activities started to shift in 1965 from New York to Dover; by 1969 there was no New York involvement.

ILC and International Latex separated finally after Apollo. When all operations were moved to ILC's Moonwalker Road facility in Frederica, Delaware the physical separation was complete. International Latex sold off its remaining interest in 1984, thus allowing what is now ILC-Dover LP (Limited Partnership) to continue its journey in supporting human space exploration.

FINAL STAGES OF HAMILTON'S APOLLO 11 BACKPACK

In 1966 the Apollo backpack life support system had no user-visible displays or controls. Such items had not been envisioned at the outset. Evaluations in 1965 revealed that a strong astronaut desire had developed for such features. In 1967 a chest-mounted display and control system called the Remote Control Unit became a parallel activity with flight certification material changes driven by flammability and toxicity issues as a result of the Apollo capsule fire. These were made parallel tasks so any delay in control or display developments would not preclude timely incorporation of the material changes. Much to the delight of the Apollo astronauts, Hamilton completed all certification activities in time for Apollo 9.

The rivalry between Hamilton and ILC led to Hamilton's Suit Lab manufacturing the outer coverings of the Thermal Meteoroid Garment for the backpack life support system concurrent with making complete spacesuits for the U.S. Air Force's Manned Orbiting Laboratory program.

The resulting backpack life support system (Figure 7.6) was the world's first,

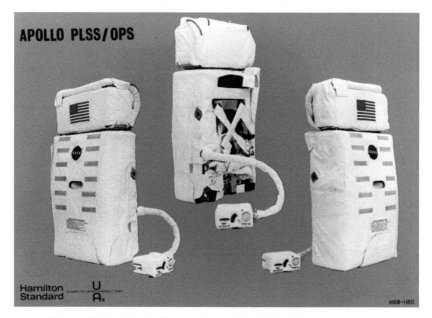

Figure 7.6. The Apollo 9 and 10 Backpack Life Support System
(courtesy UTC Aerospace Systems)

self-contained portable life support system suitable for space use. The unit furnished suited crewmen with their life support requirements, communications, telemetry, and controls and displays for lunar exploration missions.

NASA'S COMPLETION OF THE APOLLO 11 SPACESUIT

All the seemingly minor material changes introduced opportunities for unforeseen problems. NASA had to assure every element of the suit system came together and functioned well without surprises. In 1965 NASA elected to develop a radio and biomedical monitoring system for the Apollo suits. In the final push for the suits that would be used on the Moon the biomedical monitoring system worn inside the suit was again revised. This system included sensors taped to the user's skin, a harness for the electronics, and a urine collection system developed and produced by the ILC which was mounted externally to the Liquid Cooling Garment (Figure 7.7).

In parallel with the backpack for Apollo 9, ILC completed certification of the extravehicular activity accessories, which included a visor assembly, gloves with removable cover garments, a chest dust cover, and lunar overboots.

The versatile performance of the Apollo spacesuit is attributable to it being a modular system, hence assembly was required. All in all, some 18 actions were required before doing a (vacuum) "chamber run" or going for a spacewalk.

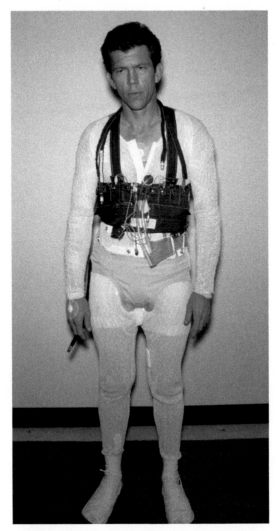

Figure 7.7. "Under suit" spacesuit elements
(courtesy NASA)

Placed in order of normal donning, these were:

1. The Biomed Belt, which provided astronaut medical data to ground control during EVA.
2. The Urine Collection and Transfer Assembly, which was a male-only devise that provided for the hygienic collection, storage, and eventual transfer of liquid waste.
3. The Fecal Containment System, which provided for emergency containment of solid waste matter when spacecraft facilities were not an option.

4. The Constant Wear Garment, which provided abrasion protection and better humidity control for the Command Module Pilot; or the Liquid Cooling Garment, which provided crew cooling in Lunar Module spacesuits.
5. Astronaut wristlets, which provided the wrists with abrasion protection and greater comfort.
6. The Pressure Suit Assembly, which involved Lunar Module crewmembers connecting the cooling garment to the suit as part of the entry process.
7. Entry zippers, which had to be closed.
8. The Communications Carrier Assembly or "Snoopy Cap," which supplied communications and bump protection to the astronaut's head, had to be put in place and adjusted.
9. The Oxygen Purge System (backup life support module), which had to be placed on to the Potable Life Support System a.k.a. "backpack."
10. The Remote Control Unit, which was then attached to the backpack.
11. The Lunar Overboots, which were put on over the suit's pressure boots.
12. The Backpack Assembly, which was attached by straps to the pressure suit.
13. The Remote Control Unit, which was attached to the chest of the suit for easy access to controls and displays.
14. The Life Support Hoses, which were connected to the front of the pressure suit.
15. The Lunar Extravehicular Visor Assembly, which was placed over the pressure suit's bubble helmet for eye protection from permanently blinding light and reflections that could obscure vision and for thermal protection.
16. The Bubble Helmet (with visor assembly), which had to be attached to the suit. The helmet provided a feed port for food and drink while being pressurized.
17. Donning Comfort Before Gloves, which were optionally donned not just for abrasion protection but also better thermal and moisture control to the hands.
18. The Extravehicular Gloves, which were donned last. The Apollo astronaut was now ready to go for a spacewalk. These EV gloves featured a Velcro attachment of the thermal overglove to a pressure glove underneath. This thermal overglove could be easily removed to provide a spare pair of pressure gloves for in-flight emergencies or reentry.

This spacesuit was a technical marvel for its time and drew a great deal of attention. Figure 7.8 shows NASA dignitaries demonstrating an Apollo 9-type spacesuit to President Lyndon Baines Johnson.

Of course, much more was involved in the Apollo 9 effort than certifications. Flight hardware had to be assembled, tested (Figure 7.9), and delivered. Then, extensive, complete spacesuit evaluations had to be performed in a vacuum chamber (Figure 7.10) to assure the astronaut was entirely familiar with the spacesuit's operations and that everything was working properly. Only then could the spacesuits be packaged for shipment to Cape Canaveral and loaded into the rocket that would put the astronauts into Earth orbit so that they

Figure 7.8. A Presidential Apollo spacesuit demonstration
(courtesy NASA)

Figure 7.9. Testing an Apollo 9 backpack before manned use
(courtesy UTC Aerospace Systems)

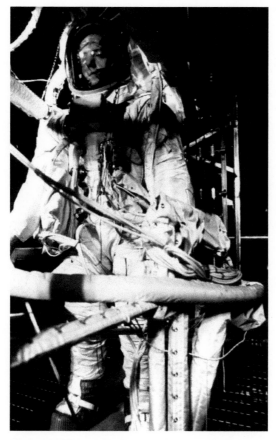

Figure 7.10. Apollo 9 Pre-Flight Chamber testing
(courtesy UTC Aerospace Systems)

could carry out orbital verification of the Apollo spacesuit before treading on the Moon.

Apollo 9 launched into Earth orbit on March 3, 1969. The primary purpose of the mission was to demonstrate the ability to perform a contingency space transfer of astronauts from the Apollo Lunar Module to the Command Module should there be difficulties docking during the return rendezvous. Additionally, this was to validate that the suits and backpacks could support the Apollo spacewalks.

The training leading up to this mission paid off. Unlike the Gemini spacewalks, the Apollo ventures into the vacuum of space were carried out seemingly easily and the mission goals were accomplished. The Apollo spacesuits always brought the astronauts back into the safety of the spacecraft without loss of life or termination of the mission.

The spacesuits appeared to perform flawlessly (Figures 7.11 and 7.12). The only post-flight comments relating to the suits were reflected light glare within

Figure 7.11. Apollo 9 testing of Lunar Spacesuit
(courtesy NASA)

the helmets and some minor condensation that developed in a helmet at one point of the mission. These were addressed by an ILC and Air-Lock visor assembly redesign before Apollo 11. Apollo 9 experienced another problem that would affect the spacesuit. This was that a disturbingly large portion of the pictures were of poor quality (i.e., "blurry") as a result of camera movement while taking the picture. The events leading up to spaceflight photography give an interesting insight into this difficulty.

NASA's space cameras in the 1960s were not made in the U.S. They were manufactured by Hasselblad AB of Sweden, one of the best and most expensive camera makers in the world. The first "space camera" was purchased by Astronaut Walter M. Schirra from a camera shop in Houston, Texas. NASA then modified his camera to make it lighter. It was then used by Schirra on his Mercury flight on October 3, 1962. For Gemini, Hasselblad created a special model. Unfortunately, taking pictures with insulated pressure suit gloves proved difficult. Most of the pictures from the Gemini missions suffered from the camera moving while taking the picture. For Apollo, Hasselblad thought they had found the solution with a specially designed version of its commercial motorized camera which reduced the work required in pointing the camera and

Figure 7.12. Apollo 9 testing of Lunar Spacesuit
(courtesy NASA)

pushing the button. What had not been understood before Apollo 9 was that holding a camera still while pushing the button would prove a serious difficulty in space.

While Apollo 10 was preparing for its historic flight to the Moon, NASA expected the lessons learned from the Apollo 9 mission to be incorporated into the spacesuit that would support the early Apollo lunar surface missions. This involved changes to the Lunar Extravehicular Visor Assembly and the Remote Control Unit. The contractors had fewer than 12 weeks to redesign, build, certify, and deliver flight hardware.

The people at ILC and Air-Lock responded by adding opaque side shades and an outer shell that was covered by the Integrated Thermal Meteoroid Garment. Prototype Lunar Extravehicular Visor Assemblies arrived in time for Apollo 11 training (Figure 7.13). The completion of certification and the delivery of flight units were accomplished to successfully support the flight. This visor assembly was also used on Apollo 12 and 13.

NASA wanted improvements in the Remote Control Unit for Apollo 11. The chest-mounted control and display unit was slightly enlarged. The controls were revised for easier use with pressure gloves and the displays gained an adjustable brightness control. This should have ended the life support portion

Figure 7.13. Training for Apollo 11
(courtesy NASA)

of corrective actions. However, blurry Apollo 9 pictures added another element to Remote Control Unit activities.

A decision was taken in the spacesuit community to create a camera-holding devise that attached to the spacesuit. Hamilton Standard engineers were then given the challenge of creating the devise. As the Remote Control Unit was already being redesigned, a front bayonet-type clip was added to the center front of the unit. The clip allowed the attachment of a bracket, which in turn held the camera assembly. The spacesuit now had a new use; not only did it allow a crewmember to venture out into the vacuum of space to perform work comfortably and safely, but it also served as a camera-holding devise. The success of the spacesuit becoming a photography platform can clearly be seen in the magnificent pictures that resulted from subsequent Apollo missions.

In May 1969 the crew of Apollo 10 became the second group of Americans to orbit the Moon. Unlike Apollo 8, Apollo 10 had a functioning Lunar Module. While no spacewalks had been planned for the mission, two Oxygen Purge System modules and one main life support backpack were stowed in a rack in the Lunar Module (Figure 7.14). The exposed sides of these life support modules were covered in Velcro pile during Apollo program development because the rack happened to be on the cabin wall opposite the Lunar Module's windows. At the same time the boots that crewmembers wore in flight were given pads of Velcro hook. The purpose was that a crewmember could attach

Figure 7.14. Stowed spacesuit life support system modules
(courtesy NASA)

his feet to the backpack Velcro for stability in the zero gravity of space should celestial navigation be required in an emergency. While no celestial navigation was required, the Velcro allowed the crew to periodically stargaze in comfort for hours on the long journey.

CHAPTER 8

Humankind's First Footsteps on the Moon

It was 8 years, 1 month, and 21 days since an American President announced the goal of going to the Moon, and everything was in place and ready. It was the morning of July 16, 1969. Astronauts Neil A. Armstrong, Michael Collins, and Edwin "Buzz" E. Aldrin Jr. were all suited up in a ready-room at the Cape Canaveral launch pad. Sealed in their pressure suits the crew of Apollo 11 relaxed in recliners while prebreathing pure oxygen to purge their body tissues of nitrogen. This was important because a few hours later they marched past an army of newscasters (Figure 8.1) to a van, were driven to the launch pad, climbed into a spacecraft, and were launched into space where the atmosphere in the capsule soon decompressed from that at sea level of 14.7 pounds per square inch (psi, 100 kPa) to a 5-psi (27.6-kPa) pure oxygen cabin environment. Without lengthy prebreathing of pure oxygen before the launch, they would have developed decompression sickness.

Apollo 11 headed to a landing site in the Moon's Sea of Tranquility to accomplish the national objective. While Earthly observations had characterized the site as relatively flat and smooth, the automated landing on July 20, 1969 had to be interrupted by Neil Armstrong to avoid crashing into a field of boulders. Armstrong then flew the Lunar Module manually over the surface of the Moon almost to the point of having to abort as a result of running low of fuel before reaching an adequate landing site. However, the landing was successful. Two hours after touchdown, four and a half hours earlier than planned, Astronauts Armstrong and Aldrin were given permission to initiate preparation for the first lunar spacewalk (Figure 8.2).

Armstrong provided documentation of the post-landing condition of the Lunar Module by taking photographs of the vehicle and an iconic photograph of Aldrin (Figure 8.3). He then collected soil samples using a sample bag on a stick before he removed the television camera from the Modular Equipment Stowage Assembly and did a panoramic sweep of the Earth. Next, he mounted the camera on a tripod 40 feet (12 meters) from the Lunar Module. Aldrin joined Armstrong in trying out methods for moving around, which included two-footed kangaroo hops or what was characterized as the "lunar lope." The weight of the life support backpack resulted in a tendency to fall backward, but

Figure 8.1. The Apollo 11 crew proceeding to launch
(courtesy NASA)

this did not pose a serious problem for maintaining balance. Loping quickly became the preferred method of lunar movement. The fine surface soil was quite slippery but under the surface the ground was very hard.

On Earth the lunar lope was viewed by many around the world watching their television sets as normal for moving on the Moon and by others as the astronauts having fun, which clearly they were.

However, it was a cause of concern to ILC Reliability Engineer Dave Slack. He watched the movement and concluded the astronaut was using the pressure suit essentially as a pogo stick in that a compressive pressure load was caused to get extra spring in each stride. ILC had not designed the suit with that in mind. The next day Slack reviewed the design numbers. This did not take long as the numbers indicated the weak link was the crotch area. While he did not know for sure what the actual loading might be in the lunar lope, he was comfortable that the design had enough margin and that all was safe.

Figure 8.2. "One small step for a man"
(courtesy NASA)

Hamilton Mechanical Design Engineer Earl Bahl also regarded the lope with apprehension. He felt the astronaut was moving across the surface faster than the design requirements. If the astronaut lost his balance and fell, the life support system could be damaged in the fall with potentially life-threatening consequences. Others at Hamilton had similar fears. Hamilton recommended NASA allow a recertification to new, higher impact loads. However, NASA trusted the design and in any event there was little in the budget for additional testing, which would have been expensive.

However, not hearing formally from ILC probably begged the question and NASA formally asked ILC to carry out a lunar lope analysis. ILC provided a presentation outlining why there was no concern. Between Apollo 11 and 14 astronauts did lose their balance and fall, but the spacesuits always remained functional. Nevertheless, this did not put to rest everyone's concerns. After Apollo 14, the Apollo 7 through 14-type pressure suits had become obsolete. NASA provided a suit to ILC for destructive testing, which was a quick and inexpensive test. The crotch was the first area to fail, but it failed at such a high load that everyone felt assured there were no issues.

Figure 8.3. The Apollo 11 through 13 spacesuit in use
(courtesy NASA)

Back on the spacewalk the two astronauts inserted the pole carrying the American flag into the ground (Figure 8.4). The EVA was temporarily halted for a call from U.S. President Richard Nixon.

The suit's insulation effectively protected the lunar explorers from the intense heat of direct sunlight and the cold of the lunar shadows. Aldrin remarked that moving from sunlight into Eagle's shadow produced no temperature change inside the suit.

The astronauts later deployed the Early Apollo Scientific Experiment Package (Figure 8.5), which included a passive seismograph and a laser ranging retro-reflector. Armstrong then loped about 400 feet (120 m) from the Lunar Module to take photographs from the rim of East Crater. Aldrin collected two core samples after using a geological hammer to pound the collection tubes into the ground.

Aldrin reentered Eagle first. The astronauts had difficulty lifting a film container and two boxes containing lunar surface material samples into the Lunar Module even though they were provided with a cable pulley device. Once all was onboard the Lunar Module, the astronauts discarded their life support backpacks, lunar overshoes, and other unneeded equipment onto the lunar surface before closing the hatch to conclude the 2-hour, 32-minute space-walk. Eagle's ascent rockets were fired to send the ascent stage back into lunar orbit while leaving the descent stage on the lunar surface. The Lunar and

Figure 8.4. A demonstration of national pride
(courtesy NASA)

Figure 8.5. Early Apollo Scientific Experiment Package deployed
(courtesy NASA)

Figure 8.6. Hamilton Standard engineers watching the Apollo 11 landing
(courtesy UTC Aerospace Systems)

Command Modules then reunited in orbit for the return back to Earth. For the return the pressure suits served as part of the crew's survival and escape system. While those who had supported Apollo were confident that the crew would return safely, all watched the crew's successful return on July 24, 1969 with some trepidation (Figure 8.6).

On November 14, 1969 Apollo 12 rocketed away from the Earth on a journey to the Moon's Surveyor Crater. The location was so named because it was the landing site of the Surveyor 3 unmanned spacecraft. One of the mission goals was to see whether Apollo could allow landing close to a specific target on the Moon. When coming in to land, visual sighting of the lunar surface was lost as a result of dust being kicked up by the descent engine; but, in spite of this, the Lunar Module landed within 600 feet (180 m) of Surveyor 3.

Two spacewalks were undertaken from Apollo 12 which would have gone flawlessly had the camera not failed to operate. On Earth, newscasters scrambled to provide alternative mission coverage in simulations and commentary, which clearly was much less interesting than the actual mission. After the mission, it was learned that the camera had been inadvertently pointed at the Sun at one point during setup. Direct sunlight had burned out a tube in the camera. Astronauts Charles Conrad and Alan Bean performed two spacewalks totaling 7 hours and 27 minutes. Upon their return the astronauts reported hand fatigue from fighting the internal pressure of the gloves. The thermal outer-garments of the suits were severely worn by abrasion from lunar dust. This received little attention outside of NASA and ILC. The post-flight pictures gave the impression the exploration went smoothly. Perhaps as a result of no live-lunar images, the second lunar surface mission garnered limited public interest.

In an attempt to boost public interest, NASA played down the belief that the number 13 was unlucky in the events leading up to the launch of Apollo 13. For all the Apollo missions the spacesuit was considered key to most emergency responses. As Hamilton was also the provider of the Lunar Module's life support system and backup guidance system, NASA funded Hamilton Standard to have at least two engineers in a conference room monitoring flight communications 24 hours a day from Apollo launch to landing.

Apollo 13 rocketed into space heading for the Moon's Fra Mauro Plain on April 11, 1970. On the evening of April 13, 1970 a young Hamilton engineer named Ted Jansen was undertaking mission support in Windsor Locks, Connecticut. He was struggling to stay awake while listening to seemingly endless but routine mission chatter. He awoke with a start when he heard the words "Houston, we've had a problem." He knew the world was listening to Apollo radio transmissions and "problem" was one of many words that were forbidden in Apollo communications. As the other engineer tried to understand the nature of the emergency, Jansen called Andy Hoffman, who was then the Hamilton Space Business Engineering Manager. While understanding Apollo 13's plight was minutes if not hours away, Hoffman's reaction to use of the word "problem" was to immediately order more than 100 Hamilton engineers into the plant for support.

Apollo 13 was 205,000 miles from Earth. There was no pressure reading from oxygen tank no. 2. Simultaneously, the entire NASA and contractor network was being activated. Thirteen minutes later, Astronaut Jim Lovell looked out of the spacecraft and announced, "We are venting something ... a gas of some sort." The pressure in the remaining two oxygen tanks began to fall. Fuel cell no. 1 and no. 3 were shut down in an unsuccessful attempt to stop the leakage. The remaining oxygen kept fuel cell no. 2 operational for 15 minutes. The batteries in the Command Module were designed to provide only enough electricity for an emergency landing. The Command Module was ordered to be shut down and the Lunar Module was immediately activated to provide a "life boat." The Lunar Module was designed to keep two astronauts alive for three days. Apollo 13 had three astronauts and they were a week away from Earth reentry according to the flight plan.

To conserve electrical power, only essential Lunar Module systems were powered and even then only when necessary. Heating the spacecraft was not an option. The crew were ordered into their pressure suits to conserve body heat and stay warm. In a matter of hours the interior of the spacecraft cooled to subfreezing temperatures. Garments that had been carefully crafted by ILC personnel in Dover became layers of clothing that protected the Astronauts from hypothermia as the rescue slowly unfolded.

The life support systems of the Command and Lunar Modules had been provided by two different contractors. The carbon dioxide removal cartridges were of different sizes and shapes. Fortunately, Hamilton was the life support system provider of both the spacesuit and the Lunar Module. Hamilton had designed the carbon dioxide removal cartridges on the suit's backpack to additionally fit the Lunar Module's life support system. This interchangeability provided an immediate spare cartridge supply. While this added perhaps a day of precious time, it was not enough. NASA used an emergency, low-level, lunar orbit to reduce the return flight to four days. However, adaptation of the Command Module's carbon dioxide removal cartridge to the Lunar Module's life support system was still needed.

During Apollo, Hamilton paid one engineer per launch to travel to Cape Canaveral, now the Kennedy Space Center, to be a mission "observer." This was normally a non-working assignment that was really a reward for working long hours and outstanding performance. However, it was understood that if an emergency arose the employee was there to help. For Apollo 13 the Hamilton observer was a systems engineer named Richard "Dick" Wilde. He was a master at calculating how long support functions and other resources could be made to last. He knew the Lunar Module's life support systems as well as anyone. Wilde was also the author of the quick reference manuals called "Mini-Databooks" for the spacesuit's backpack and Lunar Module's life support system that every NASA and Hamilton EMU engineer always had at hand. He supported the hands-on creation and prove-out of the Apollo 13 carbon dioxide removal cartridge adaptation on Earth before it was needed by the crew in space.

The Apollo 13 crew successfully splashed down off the coast of Samoa in the Pacific Ocean at 12:07 PM on April 17, 1970, which illustrated the great skills of the U.S. space community and its ability to work together.

For the next flight, Apollo 14, the crew asked for the longer duration and more mobile spacesuit being planned for Apollo 15 and subsequent flights. As the mission commander was Alan Shepard, the first American to travel into space, this received considerable consideration by NASA. However, this came at a time when NASA was under considerable budgetary pressure. The mission planning compromise was that the Apollo 14 crew would be provided the new, longer duration backpack and the improved visor assembly. The remainder of the pressure suit would be the Apollo 11 through 13 configuration. Later, budgetary issues caused NASA management to rethink the backpack and Apollo 14 launched with Apollo 11 through 13-type units. This happened so late in the program that decades later one NASA manager was insistent that Apollo 14 had used the long-duration "Rover" backpacks. Only after reviewing photos of Apollo 14 backpacks discarded on the lunar surface did he relent.

Apollo 14 launched on January 31, 1971 destined for Fra Mauro, which was originally the landing site of the aborted Apollo 13 mission. Apollo 14 Astronauts Alan Shepard and Edgar Mitchell performed two lunar spacewalks. On the first EVA the astronauts collected 42.9 pounds (19.5 kg) of geological samples, deployed the TV camera and an S-band dish antenna, and planted a U.S. flag. Then the astronauts set up the Advanced Lunar Science Experiment Package about 495 feet (150 m) west of the Lunar Module and a laser ranging retro-reflector approximately 100 feet (30 m) further beyond. The spacewalk lasted 4 hours 49 minutes and traversed about 1,815 ft (550 m) of lunar terrain.

The second spacewalk was a quest to reach the rim of Cone Crater, which was formed by a meteoroid impact that had created a pit 1,000 feet (300 m) wide. Geologists hoped to find a resulting natural excavation that would reveal eons of lunar geological history, possibly including the Moon's oldest rocks. To do this the astronauts had a two-wheeled "rickshaw" cart named the Modularized Equipment Transporter for hauling tools, equipment, and samples. While the journey started with easy terrain (Figure 8.7), the path needed to reach the objective soon became rocky and the astronauts ended up having to lift up and carry the loaded cart to continue.

Even without visual distance references provided by the Earth, the crater initially appeared reachable. However, the astronauts reached the point in the mission timeline that required them to abort the walk and return to the Lunar

Figure 8.7. The easy portion of the second Apollo 14 spacewalk
(courtesy NASA)

Module before reaching the rim of the crater. While this certainly was a disappointment, the 4-hour, 46-minute walk garnered an additional 51.7 pounds (23.5 kg) of lunar samples.

After returning to the Lunar Module, Shepard and Mitchell started their close-out activities. During a delay in these activities, Shepard fastened an item that he had carried from Earth onto a mission tool handle that had already served its mission purpose. He then drew from his pocket another personally carried item and announced "Houston, while you're looking that up, you might recognize what I have in my hand as the handle for the contingency sample return; it just so happens to have a genuine six iron on the bottom of it. In my left hand, I have a little white pellet (a golf ball) that's familiar to millions of Americans. I'll drop it down. Unfortunately, the suit is so stiff, I can't do this with two hands, but I'm going to try a little sand-trap shot here." The swing missed. Shepard repeated the action. The second swing hit lunar soil. Mitchell added "You got more dirt than ball that time." The ball moved only 2 or 3 feet. However, on Shepard's third attempt he finally connected the ball but the trajectory was low. Shepard tried with a second ball, connected, and announced "Miles and miles and miles" as the golf ball sped off in the Moon's one-sixth Earth gravity. While this was not a NASA-sanctioned experiment, it did earn the first U.S. astronaut to travel into space the additional distinction of being the first human to play golf on the Moon. Shepard's homemade golf club head is currently on display at the U.S. Golf Association Hall of Fame in New Jersey.

CHAPTER 9

The Paths to the Apollo 15 through 17 Spacesuits

In October 1964 NASA set out the path they would follow to create the spacesuit for the later Apollo missions that were to include Lunar Rovers. In 1964 that path and funding source was to be the Apollo Block III program. NASA became the Apollo integrator of all Apollo spacesuit efforts when it assumed the program leadership role of Block II in 1965. However, the events of 1965 had turned the Hamilton Standard Division of United Aircraft Corporation (a.k.a. "Hamilton") and International Latex Corporation (ILC) into bitter rivals. Before the contractual issues of the original Apollo program had been resolved and the Apollo Block II contracts had been signed in March 1966, Hamilton personnel were planning a Hamilton-ILC rematch where Hamilton hoped to emerge victorious. While this would not happen, it triggered events resulting in a new and separate Apollo suit direction that culminated in the spacesuit used in Apollo 15 through 17 surface missions.

THE "ROVER" PRESSURE SUIT, BORN IN RIVALRY

Before 1965 any rivalry between Hamilton and ILC was probably the last thing Hamilton management wanted. However, in the first two months of 1965, Hamilton's President became convinced that ILC could not or would not support Hamilton's spacesuit contract and that elements within NASA's Crew Systems had lost faith in ILC's capabilities. With NASA's permission, Hamilton replaced ILC with Goodrich as the Block II suit provider. ILC came back to win the Block II suit design competition and successfully used that victory to regain the suit side of the contract and remove Hamilton as the suit system integrator. This left bitter feelings. This also left Hamilton with a pressure suit capability whenever there was another U.S. spacesuit competition in the near future.

In 1963 the Kennedy administration tasked the Air Force with the development of a space station. This was named the Manned Orbiting Laboratory or "MOL" program (Figure 9.1).

Figure 9.1. MOL Launch Vehicle
(courtesy UTC Aerospace Systems)

Hamilton had already seen MOL as yet another space business opportunity and in September 1964 had invited Air Force MOL representatives to Windsor Locks, Connecticut to review Hamilton's capabilities. At this point, Hamilton was proposing to the Air Force that ILC would be its pressure suit provider should Hamilton win any MOL spacesuit work.

During the visit, Hamilton learned that the Air Force was having great difficulty gaining biomedical data from chimpanzee test subjects. The medical sensors were uncomfortable and the chimps easily removed them with their hands or their feet. Furthermore, the Air Force had no means of collecting carbon dioxide and exhalation humidity data as chimps would not wear face masks. Hamilton's answer was the Chimp Suit (Figure 9.2).

The suit system essentially consisted of a helmet sized and shaped to fit a chimpanzee with a simple fabric coverall-type torso assembly to help hold the helmet in place on the suit subject. The helmet had accommodations for inlet and outlet hoses. The helmet prevented leaking the atmosphere by way of a soft rubber seal around the neck. The torso assembly used synthetic padding material on the inside of the chest area, padded thumbless mittens, and legs that linked together at the feet as part of a leg restraint system to discourage the chimps from removing the uncomfortable biomedical monitors. The Chimp Suit also included a compact closed-loop life support system that had sensors to capture in-suit atmospheric conditions. This allowed measurement of the chimp's oxygen consumption and exhalation by-products.

As an encouragement for ILC to meet the technical challenges in the Apollo program and to reaffirm ILC as Hamilton's suit MOL fabricator, Hamilton funded ILC to fabricate the torso assembly for the Chimp Suit even though

Figure 9.2. The restored "Chimp Suit"
(courtesy UTC Aerospace Systems)

Hamilton had by that time a "Suit Lab" and an internal capability for pressure suit manufacturing. ILC delivered the torso assembly at the end of December 1964. At the beginning of January 1965 relations between Hamilton and ILC began to break down into an unworkable impasse. Hamilton elected to delay the delivery of the Chimp Suit. After Hamilton changed its Apollo program pressure suit fabricator in March 1965, Hamilton funded the internal manufacture of a second iteration of the torso garment, assembled the system with the Hamilton-made torso, and delivered it to the Air Force for evaluation. The ILC-made torso remained part of a pile of sewing examples at Hamilton for almost 40 years before being restored to be part of an historical display (Figure 9.2).

To support Apollo pressure suit development, Hamilton had internally funded the creation of a "Suit Lab" with sewing (Figure 9.3), layout, and dedicated suit-testing areas. With the Apollo-related B.F. Goodrich personnel gone and no external funding for suit development, this left considerable factory

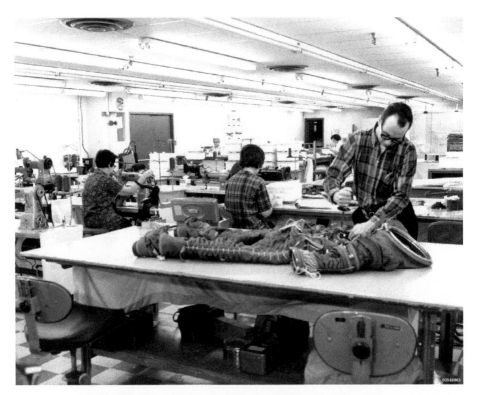

Figure 9.3. Hamilton Standard's Suit Lab sewing area
(courtesy UTC Aerospace Systems)

space idle. Even as Hamilton was in the process of formally surrendering responsibility for the pressure suit portion of the Apollo spacesuit in late 1965, it was already working on defeating ILC (and the David Clark Company) in the upcoming MOL spacesuit competition and on a strategy to win back the pressure suit portion of the Apollo spacesuit. By the time Hamilton had contractually surrendered the suit system integration and pressure suit management roles in March 1966, it had already internally funded five successive prototype pressure suits for its parallel MOL program.

Hamilton had an advantage in that it was required to conduct suit system-level testing for NASA. Testing their life support system on the current Apollo suits allowed Hamilton personnel to also gain firsthand knowledge of the ILC suit's strengths and potential shortcomings. Hamilton felt the ILC suit had three areas of weakness although they also recognized that their five MOL prototypes all had areas of performance where the ILC suit was better.

The first ILC area of weakness was that the suit did not easily permit bending over and picking up objects from the surface. The second weakness was that the torso section became round as a result of pressure increasing the front-to-back volume. When the wearer used the front draw strap to force the suit into a bent-over position to operate a Rover, the neck ring also rose as the shape changed. The result was that the person tended to sink into the suit causing his head to drop partially out of the helmet, which limited vision. The third weakness was that most users could not self-don the suit.

By this time Hamilton management had fully taken on board the events of 1965. They reasoned that the shoulders, arms, and Durney brief section of the ILC Apollo suit had been developed under contract to Hamilton, so they felt they had legal rights to use those features.

In March 1966 Hamilton management provided internal funding to build a Hamilton replica of the ILC suit but using Hamilton fabric convolutes in place of ILC's molded rubber. The task was assigned to Mark Baker who delegated it to John Korabowski, Mike Marroni, and Doug Gettchell. The team elected to make some "improvements." As directed, their design used two layers of fabric, one a restraint layer, the other a coated pressure barrier, rather than molded rubber in the joints. However, the suit used a different entry system, which allowed it to have a multidirectional waist joint as well.

The resulting suit (Figure 9.4) proved to be lighter, more mobile, and better able to operate in confined spaces. Hamilton decided to use this as their competition suit for the upcoming MOL contract and only released pictures of the suit with cover garments attached to avoid prematurely disclosing any technical details of the design (Figures 9.5 and 9.6).

With an eye to a possible competition for Apollo Block III, the Hamilton MOL prototype was designed around a 5.0-psi (34.5-kPa) Apollo Block III operating pressure rather than the 3.7-psi (25.5-kPa) pressure requirement for Apollo Block II and MOL. This was expected to provide greater durability and mobility at the lower Apollo Block II and MOL operating pressure.

Moreover, the Hamilton MOL suit provided better visibility than the ILC Apollo suit when seated (Figure 9.6). Hamilton management were so pleased with the results that they immediately funded building a second Apollo version that could connect with the Apollo backpack. While this suit was completed in July 1966, demonstrating it to NASA took second place to completing the MOL prototype where Hamilton won against a field that included the David Clark Company, ILC, and Litton.

Apollo testing indicated that zero gravity was the hardest environment in which to don the ILC Apollo suit. At the beginning of 1967 the only way to realistically simulate zero gravity on Earth was to fly an airplane up to as high an altitude as possible and then go into a precisely controlled dive where the plane is dropping at exactly the same rate as the pull of gravity. This provided zero gravity for less than half a minute before the simulation had to be stopped to allow the airplane to end the dive and regain altitude. Lunar gravity can be simulated in the same manner but with the airplane dropping at a slower rate. By this time NASA was also experimenting with another way to simulate zero gravity on Earth. It was under water in a swimming pool. However, in that environment the resistance of water made donning a less exact replication. To don the suit the suit subject's body had to push water out of the suit in order to

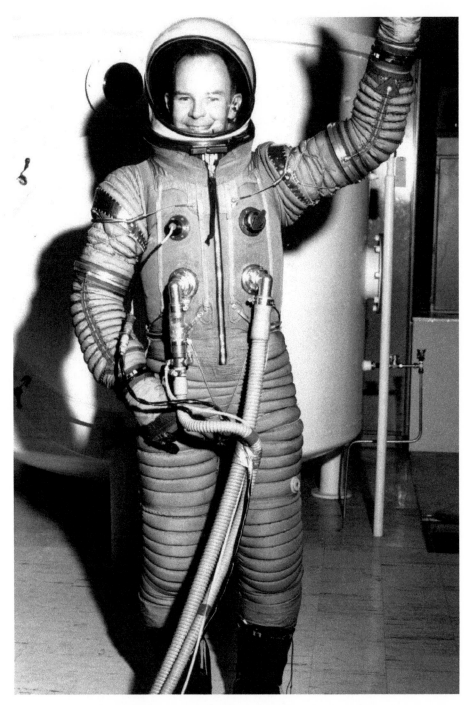

Figure 9.4. Hamilton MOL Competition Suit less covers
(courtesy UT Aerospace Systems)

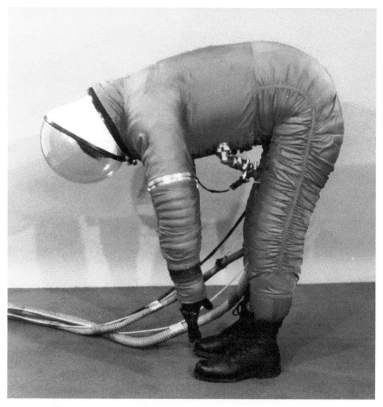

Figure 9.5. Hamilton MOL Competition Suit bending
(courtesy UTC Aerospace Systems).

enter. This made the process much more difficult. This inspired Hamilton's MOL Manager Jack Kelly and its (space) Business Development Unit.

In February 1967 Kelly performed a zero-g, unassisted don/doff demonstration. The "neutral buoyancy facility" Hamilton used in the demonstration was the swimming pool at Central Connecticut State College in New Britain, Connecticut. This highlighted the advantage the Hamilton suit had over the Apollo program's suit in that all the entry components of the Hamilton suit were within easy reach of the crewmember. Kelly was filmed sliding in and out of the Hamilton suit with ease (Figure 9.7).

While the Apollo capsule fire the month before caused NASA to be almost totally focused on program recovery, NASA did elect to have the Air Force ship the Hamilton MOL competition suit to Houston for testing. The suit was tested at Apollo Block II and Block III pressures. One of the test subjects was Joe McMann.

During a treadmill running test a leg cable failed on the Hamilton MOL suit. McMann remembers it well because one leg immediately became longer and his foot was forced out of the boot. Fortunately, he instantly hopped on one foot

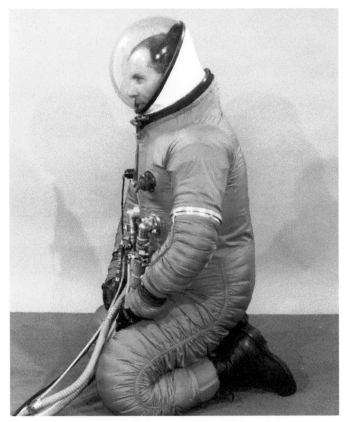

Figure 9.6. MOL Competition Suit sitting
(courtesy UTC Aerospace Systems)

so as not to trip and successfully stopped the treadmill. This did not reflect well on Hamilton. The use of "aircraft cables" in spacesuits was and still is an art. These cables consist of a plastic coating over multiple strands of thin stainless steel wire wound together. The connector ends are attached by crimping. Crimp too hard and the steel strands in the cable become damaged and fail with use. Crimp too lightly and the cable can pull out. This is a skill that Hamilton, ILC, and the Russians all had to learn. After Apollo, the U.S. elected to do away with plastic-coated steel cables in spacesuits, so as not to depend on such skills. However, the Russians and Chinese still use such cables today.

In parallel with Hamilton's proactive moves the Apollo program had continuously revised the life support system connectors of suits right back to the Gemini program making Hamilton's 1966 Apollo prototype obsolete. Hamilton had also made significant improvements in suit mobility and durability during 1967 on its MOL suits. The suits gained improved shoulder and waist designs (Figure 9.8). With Hamilton's 1967 internal funding committed to

Figure 9.7. Weightless don/doff demonstration
(courtesy UTC Aerospace Systems)

MOL-related activities, their next Apollo suit move had to wait until the start of 1968.

At the start of January 1968 Hamilton funded the manufacture of two Apollo prototypes with the same basic design that were simply named "Apollo Suits" (Figures 9.9 and 9.10). These suits were essentially MOL training suits with Apollo configuration vent systems and life support system connectors. These suits were extensively tested and photographed by Hamilton in preparation for a presentation to NASA. During a break in one of the test sessions, suit subject James "Jim" Fentress stood on his head as a joke. The photographer captured his performance (Figure 9.9). It would later be used by Hamilton's Business Development Unit to show how Hamilton personnel were ready to stand on their heads for NASA.

Hamilton testing of these suits with Apollo backpacks validated that the suits could support Rover use and more effective surface exploration (Figure 9.10). Moreover, the testing demonstrated the differences between NASA's ILC suits and the Hamilton suits made little difference to the function of existing design backpacks. So, Hamilton representatives traveled to Houston in April 1968 to offer NASA an already designed, more mobile pressure suit that met the expected Apollo Rover requirements. Hamilton found NASA unreceptive.

Figure 9.8. Richard Lawyer testing a MOL suit
(courtesy UTC Aerospace Systems)

While this appeared to some as "politics," budget and schedule challenges were real issues to NASA. Under government contracting requirements, consideration of a new suit would involve another open suit competition, selection of a winner, a lengthy contract negotiation, and an awards process before any new suits could be made. This could easily take more than a year. Another issue was funding.

NASA's budget was reduced in 1968 and the Presidential election was indicating this was a harbinger of the funding trend to come. This proved to be true when NASA headquarters informed Houston that 1969 funding would be lower. The final 1969 NASA budget was less than 72% of its 1965 funding, which was the all-time high for the Apollo era when adjusted for inflation. To switch pressure suit contractors, there would be a period where the outgoing and incoming organizations would both have support operations working simultaneously, bringing about a period in which expenses would double.

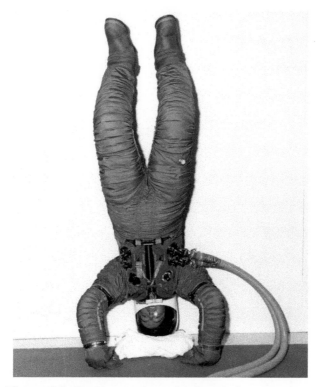

Figure 9.9. Jim Fentress in an Apollo suit demonstration
(courtesy UTC Aerospace Systems)

Adjustments in NASA's strategic planning resulted in the cancellation of the Block III development program in 1969 as well as the Apollo 18 and later lunar missions in 1970.

The Hamilton Apollo suits were never evaluated by NASA. The planned building of a more advanced Apollo prototype was immediately terminated by Hamilton. However, their efforts and presentations detailing the shortcomings of the existing ILC Apollo suits did have an effect within the NASA community.

It is currently unknown who was responsible for the name "Omega Project" (Figure 9.11). The overall idea seems to have originated at ILC and elements within NASA. The first known activity appears to have been a very informal study conducted by NASA and ILC field personnel in the summer of 1968. This study looked closely at entry concepts. This was accomplished by taking many worn-out or obsolete ILC Apollo suits and hand-sewing zippers onto the torso of the garment to represent proposed entry configurations. The ILC technicians then opened the zippers and cut the pressure garment to create a corresponding opening in the garment. This permitted evaluation of various entry concepts. NASA and ILC personnel were then able to reduce the

Figure 9.10. A 1968 Apollo Suit Picking-up A Block
(Courtesy UTC Aerospace Systems)

concepts under consideration to two. These configurations gained the names "A8L" and "A9L."

On September 10, 1968 a NASA-ILC team presented an Omega Project proposal to NASA management to build A8L and A9L prototypes so that the concepts could be competitively evaluated. Now, it is unclear whether or to

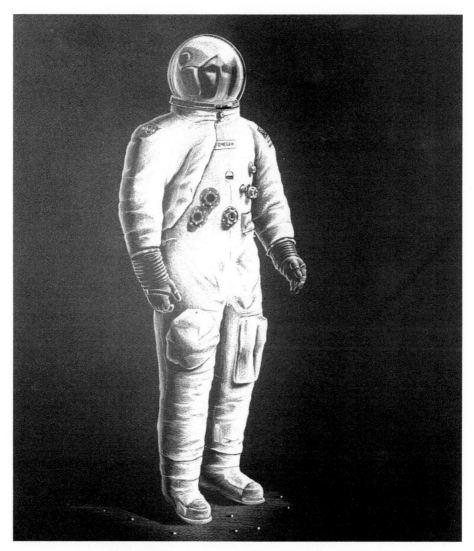

Figure 9.11. An Omega suit proposal illustration
(courtesy ILC-Dover LP)*

what level the A8L and A9L prototypes were built. Most Hamilton versions of Apollo suit history say complete A8L and A9L prototypes were made. All ILC versions say there was no prototyping until NASA directed such. This difference might be a result of confusion at Hamilton about the aforementioned effort in which zippers were hand-sewn on obsolete pressure suits during preliminary prototyping. Another possibility is that, as a result of schedule constraints, ILC may have actually built A8L and A9L prototypes in advance of being directed to do so but could not admit their existence because of contrac-

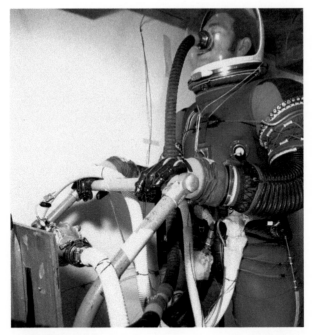

Figure 9.12. Joe McMann testing a A7LB prototype
(courtesy NASA)

tual regulations. Either way, funding was not approved by NASA. Instead, NASA management directed the NASA-ILC team to select one concept for prototyping. The selected concept was the former A9L configuration. This was subsequently renamed the A7LB to indicate that this was simply an update of the current Apollo design which could be funded under the existing ILC-Apollo suit contract without competition.

NASA's Joe McMann had the honor of testing an early A7LB prototype in a vacuum chamber (Figure 9.12). This was a metabolic test that required strenuous activity.

Unbeknown to McMann, a little pulley on the inside right thigh of the suit was not rotating properly. With every stride, the crotch cable was sawing through the pulley. Suddenly, the right leg popped and "grew." McMann had great difficulty keeping his balance until the treadmill was stopped. During repressurization of the chamber, McMann recalls that it "got really warm down there [thigh area] as the suit collapsed around me." NASA tried duplicating the failure outside of a chamber "at sea level" atmospheric pressure for safety reasons. The NASA test subject was Jim O'Kane. The suit was instrumented to measure pulley temperature. As the test progressed, the pulley temperature rose and so the test continued and continued. While NASA was unable to wear out the pulley, they did succeed in wearing out O'Kane. In subsequent attempts, NASA was never able to reproduce the pulley failure.

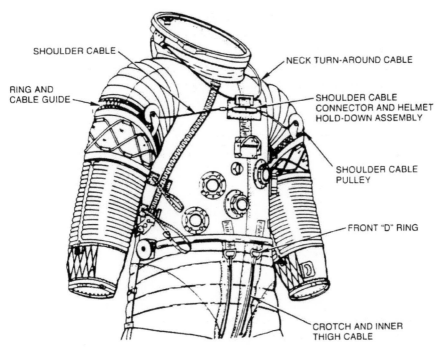

Figure 9.13. New lunar suit features
(courtesy NASA)

There were two versions of the A7LB Apollo suit. The extravehicular (EV) configuration was the "new suit" developed by NASA and ILC (Figures 9.13 to 9.15). This would be used by the astronauts who made lunar landings and walked on the Moon. These suits had a mid/side-entry system as opposed to the rear-entry suits used on Apollo 7 through 14. The Apollo 15 through 17 EV suits featured a neck joint to make it easier for the astronaut to look down when walking or forward when operating the Rover. The shoulders were revised to reduce effort and bulk. A waist joint was added and the front draw strap removed to make it easier to sit or pick up samples. The gloves were revised to reduce manufacturing costs and to enhance durability. These suits served reliably and with distinction in the Apollo 15 through Apollo-Soyuz Test Project.

In 1969 an ILC suit subject was larking about while suited in an A7LB during a break in testing. He started making like he was a football player. It inspired an outdoor film session on an actual football field where the suit subject demonstrated his ability to play football in a fully pressurized A7LB suit. NASA personnel had witnessed ILC suit subjects demonstrating greater mobility in ILC suits than NASA crewmembers or other suit testers could achieve. To explain NASA's inability to match the performance of ILC personnel, NASA jokingly spoke of "ILC gorillas." This had nothing to do with strength. The ILC suit subjects spent so much more time in the suits that they

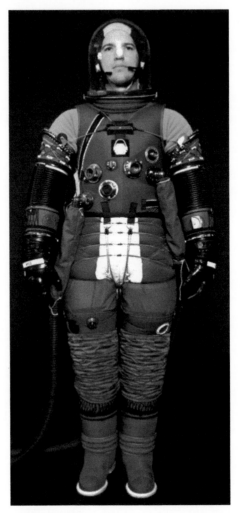

Figure 9.14. The new suit less covers
(courtesy Gary L. Harris)

had probably unknowingly learned the techniques of speed and direction to get optimum mobility with less effort.

It was also in 1969 that the Nixon Administration cancelled the Air Force's Manned Orbiting Laboratory (MOL) program. The Air Force offered NASA its inventory of MOL pressure suits. While this offered Hamilton the brief hope of supremacy in its rivalry with ILC, it occurred too late to be an opportunity. The MOL's thermal requirements were different from Apollo's and hence would have caused a total redesign and remanufacture of the MOL suits. With the loss of the MOL contract, Hamilton ceased to be a pressure suit manufacturer and competitor to ILC or David Clark.

Figure 9.15. The Apollo 15 through 17 Lunar Suit
(courtesy NASA)

CREATING THE LIFE SUPPORT SYSTEM FOR LONGER LUNAR TRAVELS AND EMERGENCY HIKES

The Apollo 11 through 14 backpack life support system was nominally certified for six hours. The backup life support system was certified for only 30 minutes. Rover missions brought up the possibility of the breakdowns and astronauts having to walk back to base camp. This meant the Apollo program needed greater life support capacity and hence further development. To that end, NASA originally asked Hamilton to retain the external envelope but produce a system that was certified for eight hours. NASA also wanted a two-hour backup life support system that could do all the functions of the primary life support system, which was the large backpack assembly. However, NASA wished this backup life support system to be still mounted on top of the primary system and fit into the same volume as the 30-minute backup system used on Apollo 9 through 14.

Evaluation of the Apollo 11 through 14 main backpack found the eight-hour capacity could be easily obtained in all but two areas: oxygen and water storage. Testing proved the primary oxygen bottle could operate at pressures higher than the original design. Thus, the increased oxygen requirement was met by charging the oxygen bottle to a higher pressure.

Increasing the amount of water stored for an extra two hours was not as simple. There was not enough room in the backpack to increase the water capacity cost-effectively. A review of the Lunar Module interfaces by Grumman and NASA showed that a protrusion on the right side of the backpack (as worn) up to 1.625 inches would not require a Lunar Module interface redesign or cause a crew mobility area constraint. A right side-mounted, auxiliary water tank, called the "Volkswagen tank," was designed and added to create the extra water capacity for the "eight-hour" backpack (Figures 9.16 and 9.17).

Obtaining a two-hour backup life support system was a much greater challenge. This required essentially quadrupling the capacity of the life support system but not changing the external size. Hamilton's engineers successfully

Figure 9.16. The Apollo 15 through 17 Backpack less covers
(courtesy UTC Aerospace Systems)

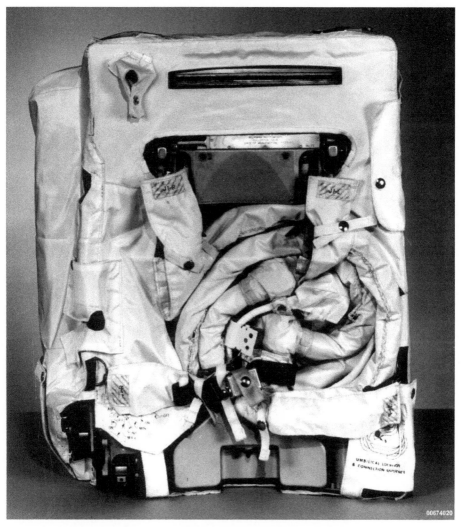

Figure 9.17. The Apollo 15 through 17 Backpack with covers
(courtesy UTC Aerospace Systems)

rose to the challenge by producing a design called the Secondary Life Support System (Figure 9.18). Unfortunately, some elements at NASA expected the physically smaller life support system to cost correspondingly less than the main backpack. Hamilton's presentation showed the cost was slightly greater than the main backpack, which was not well received at a time when the NASA budget was shrinking. This caused both NASA and Hamilton to rethink the requirements as well as the solution.

With the Oxygen Purge System (OPS) on the low-flow setting, the existing backup life support system could provide 75 minutes or more of gaseous life

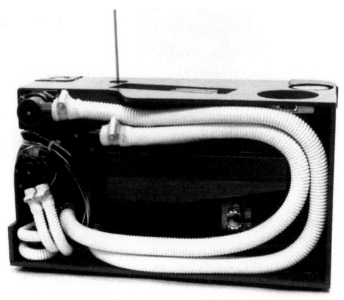

Figure 9.18. Hamilton's Secondary Life Support System prototype
(courtesy UTC Aerospace Systems)

support. However, no cooling would be provided to the astronaut. Assuming one worst-case scenario at a time and Hamilton being permitted to tweak the redesign of the main life support in progress, the resulting main life support system backpack could cool two astronauts simultaneously. If the backup life support system requirement was increased from 30 minutes to 75, then a jumper-hose umbilical system (Figure 9.19) could cool an astronaut whose backpack had failed. Upon reflection NASA found they were able to live with a 75-minute emergency capacity.

Hamilton's resulting jumper-hose system (Figure 9.20) was consequently named the Buddy Life Support System. This was carried on Apollo 14 through 17. Neither the Buddy Life Support System nor the Oxygen Purge System had to be used on any Apollo spacewalk.

Introduction of the Lunar Rover did not just affect the flight life support systems, it also affected what was needed for training. Vacuum chambers were too small and too costly to permit realistic training with a Rover replica in them. Umbilicals were too restricting to allow accurate mission simulations even outdoors. Moreover, the cryogenic ground test backpack called the Liquid Air Pack that was used from 1965 to 1969 required umbilicals for simulated radio communications and additional cooling in situations where training required strenuous activity or when training was being performed outside in hot weather.

In 1970 NASA contracted Hamilton to redesign the Liquid Air Pack to additionally provide cooling through the Liquid Cooling Garment and to give

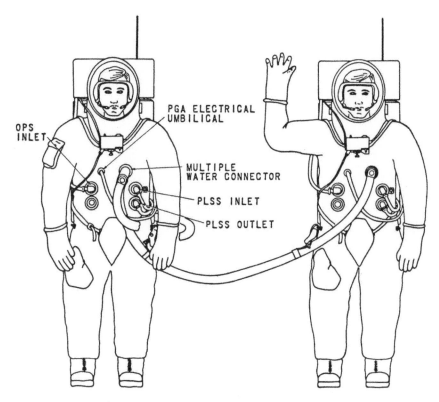

Figure 9.19. The Jumper Cooling concept
(courtesy NASA)

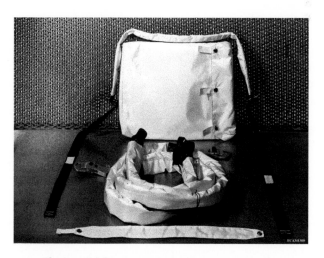

Figure 9.20. The Buddy Life Support System
(courtesy UTC Aerospace Systems)

Figure 9.21. The Improved Cryo-Pack
(courtesy NASA)

greater overall cooling capacity. This was to be done while retaining the same envelope, not significantly increasing weight, and providing all the features of the flight system such as pressurization, humidity control, gaseous life support, and communication.

Astronaut liquid cooling was accomplished by circulating water through the backpack and a Liquid Cooled Garment by a batterypowered pump. Cooling the water was accomplished in two ways. One was by vaporizing liquid air as it emerged from the storage tank. The other was by melting ice. The resulting ground-training backpack was named the Improved Cryo-Pack (Figure 9.21).

The Improved Cryo-Pack successfully served the astronauts in training for Lunar Rover-related activities (Figure 9.22) and virtually every other aspect of the Apollo 15 through 17 missions.

With improved spacesuits, ground-training equipment, and extensive training, the crews of Apollo 15, 16, and 17 were then ready to make their historic explorations of the Moon.

THE LAST THREE LUNAR SURFACE MISSIONS

Apollo 15 was the first mission to have five spacewalks: four on the lunar surface and one in deep space during the return flight. Surface explorations

Figure 9.22. The Improved Cryo-Pack supporting Apollo 15 training
(courtesy NASA)

were spent surveying the Hadley-Apennine mountain range. The Lunar Rover
Vehicle was used on the second, third, and fourth spacewalks of the mission.
The first spacewalk came shortly after landing on July 30, 1971. The landing
site was selected for its relatively flat topography in a valley. Like Apollo 14,
dust was kicked up during descent and obscured final landing visibility. To
verify having landed in the correct location and to aid in planning the surface
excursions that followed, Commander David Scott and Lunar Module Pilot
James Irwin received permission for a Stand-up Extravehicular Activity or
"SEVA." To do this, they depressurized the Lunar Module. Using umbilical
connections to the vehicle's life support system, Scott stood up with his
shoulders through the top hatch so he could get his bearings. This SEVA lasted
33 minutes.

The next day brought the second EVA, which was the first surface excursion
of the mission. Scott and Irwin started by setting up the improved TV camera
and collected a contingency sample, respectively. The astronauts then unstowed
and deployed the Lunar Rover Vehicle developed by Boeing. Irwin would lose
his balance and fall several times during the deployment without harm to
himself or his spacesuit. Although the Rover was designed to have four-wheel
steering, the astronauts found the front-wheel steering was not operating, but
they were still able to perform the mission using the rear steering. About three
hours into the sortie, Scott and Irwin set out on a 6.2-mile (10-km) exploration
heading south along the rim of Elbow Crater to St. George Crater, near Hadley
Rille (Figure 9.23). During the ride the astronauts reduced suit cooling to avoid
becoming cold since their metabolic rates were low while riding in the rover.
The astronauts encountered some difficulties from a condition called "zero-

Figure 9.23. Irwin with the Rover, Mount Hadley in background
(courtesy NASA)

phase lighting." This phenomenon occurs when light reflected from the landscape opposite the Sun is almost as bright as from the direction of the sun.
This makes obstacles difficult to discern. The astronauts used a rake to collect
"walnut-sized samples" near St. George Crater before returning. Upon their
return the astronauts deployed an Advanced Lunar Science Experiment
Package. Scott had used more oxygen than expected, so flight controllers terminated the EVA 30 minutes early. This second Apollo 15 EVA lasted 6 hours
and 34 minutes.

During this outing, Irwin had been extremely thirsty because his drink bag
failed to operate. It failed to operate during any of his Apollo 15 spacewalks.
Moreover, both astronauts suffered pain in their fingers. This was caused by
their fingernails making hard contact with the inside of the glove fingertips.
Irwin needed help to remove his gloves, and elected to trim his nails before the
next EVA. Scott left his fingernails alone out of concern for reducing touch
feedback. Dust from the EVA made the life support system connectors tight
and difficult to operate. However, this and the Rover missions to follow left
lasting tracks that prove humans traveled there (Figure 9.24).

The third Apollo 15 spacewalk came on August 1, 1971. In this 7-hour,
13-minute expedition, Scott and Irwin would utilize the Rover to cover 7.5
miles (12.5 km), explore the area southeast of Mt. Hadley Delta, and deploy a
heat flow experiment. For this excursion the front steering on the Rover
inexplicably operated. They traveled to the foot of the Hadley Delta mountain,

Figure 9.24. Tracks from the return trip
(courtesy NASA).

passing Index, Arbeit, Crescent, Spur, and Window Craters. At Spur Crater, they collected the "Genesis Rock," which today is still believed to be a piece of original lunar crust that is more than four billion years old. Scott called Spur a "gold mine" of interesting geological samples. Back at the lunar base, Scott had difficulty drilling a core hole, hurting his hands in the effort. Then the 10-ft-long (3-m-long) core tube was reluctant to be removed and was left until the next day. At the end of this 7-hour, 16-minute outing, the U.S. flag was planted as the conclusion of the day's activities.

August 2, 1971 saw Apollo 15's fourth spacewalk, which was the last lunar surface excursion of the mission. This lasted 4 hours and 20 minutes. This was Scott's fifth career spacewalk, which established a world record that would not be surpassed until 1984, when Cosmonauts Leonid Kizim and Vladimir Solovyov performed six EVAs outside Salyut 7. The walk was delayed almost two hours to let the crew rest after they experienced irregular heartbeats. This was traced later to potassium deficiency which was, in Irwin's case, complicated by failure of his suit drink bag. Scott and Irwin managed to free the core tube, which became stuck on the last outing, but could not initially take it apart to stow it because of problems with a malfunctioning vise. They finally succeeded after 28 minutes by using a wrench. Irwin and Scott proceeded west on the Rover to Scam Crater, and then turned northwest to Hadley Rille. This "drive" marked the first time Apollo astronauts passed out of sight of their Lunar Module.

The Rover permitted the Apollo 15 lunar crew to travel almost 28 miles (50 km) at an average speed of 5.7 mph (9.16 km/hr). The longest single "drive" was 14 miles (23 km) with a maximum distance from the base camp of 3 miles (5 km). During the three Rover excursions the astronauts collected nearly 176 pounds (80 kg) of samples.

Apollo 16 launched in April 1972. This was the only expedition planned to the lunar highlands. John Young was the Mission Commander. Charles Duke and Thomas Mattingly were the Lunar Module and Command Module Pilots, respectively. The mission would include four EVAs. Three would be on the surface of the Moon. Lunar dust continued to create difficulties making zippers stick, impeding the function of glove and life support system connectors, impairing the movement of bearings, as well as making it hard to read gauges. These conditions are now understood as normal for lunar exploration.

The first spacewalk came on April 21, 1972 and lasted 7 hours and 11 minutes. In preparing for this outing, Duke had trouble getting into his suit because he had grown 1.5 inches (4 cm) in the weightlessness of space. This was a physiological effect of weightlessness that the Apollo program had not taken into account during suit fitting.

Young and Duke deployed the U.S. flag and an experiments package before preparing to depart on the Rover. The Rover initially started with no rear steering indicating one battery was low on power, but became fully operational once under way. Young and Duke drove past Flag, Spook, Buster, and Plum craters collecting samples as directed by videoconference with geologists on Earth.

After their return, Mission Control relayed the news that the House of Representatives had approved initial funding of the Space Shuttle. John Young made an impressive leap and saluted the flag. This gesture would have future significance as Young went on to be the Commander of the first Shuttle mission in 1981. Duke also jumped for joy but slipped and fell on his life support system backpack causing much concern to Hamilton and NASA personnel on Earth. Fortunately, no harm came to the astronaut or his spacesuit.

On April 22, 1972 Young broke the radio antenna off his backpack while working his way out of the hatch for the second surface outing, which resulted in a small reduction in signal strength. The astronauts performed geological surveys as they traveled to and returned from Stone Mountain. The excursion lasted 7 hours and 23 minutes.

The next day's exploration goal was Smoky Mountain. Young and Duke explored safely and successfully for 5 hours and 40 minutes. During the three Rover spacewalks the astronauts traveled 16.6 miles (26.7 km) carrying tools and samples at an average rate of 4.8 mph (7.8 km/hr). The longest day's exploration distance was 7.2 miles (11.6 km) reaching a point that was 2.7 miles (4.5 km) from the Lunar Module. With the surface explorations complete, Young and Duke returned to lunar orbit and successful rendezvous with the Command Module for Apollo 16's return to Earth.

Apollo 17, the last manned mission to the Moon, had some differences from

Figure 9.25. Cernan taking a test drive
(courtesy NASA)

the preceding missions. It had a career geologist, Harrison Schmitt, as one of the lunar explorers. The television equipment was omitted to save weight and extend the Lunar Module's hover time before having to land or abort. Veteran Astronaut Eugene Cernan was the Mission Commander.

The first EVA was carried out on December 11, 1972. The first crew task involved unloading the Rover, followed by experiment deployment and planting the flag. Part of a Rover fender was accidentally damaged in the process. Cernan made a duct tape repair and did a brief checkout drive (Figure 9.25). The repair appeared successful, so the pair headed south to Steno Crater in an area called the Central Cluster. Unfortunately, the damaged fender fell off before the first scheduled stop. The astronauts were showered with dust but continued. Two explosive packages were placed so as to go off after Cernan and Schmitt departed. The resulting effects were recorded by instruments in the Apollo 17 experiments package that was left behind. The outing lasted 7 hours 12 minutes. During the crew's sleep period, John Young (Apollo 16's Commander) led the effort to develop a repair for the next day back on Earth.

Before the start of EVAs on December 12, 1972 the repair procedure for the Rover fender was transmitted to Schmitt and Cernan. The repair was accomplished using folded maps and two lamp clamps. After completing the repair the astronauts set off to conduct a geological exploration on a path that took them to South Massif. At 7 hours and 37 minutes this outing set a world record for the longest spacewalk. The record stood until May 13, 1992 when it was surpassed by three STS-49 astronauts going out simultaneously to repair a satellite.

On December 13, 1972 the last 20th-century human exploration of the Moon was conducted. This day's activities took the lunar crew to North Massif. Their EVA time was 7 hours and 16 minutes. During the three Apollo 17 sorties, astronauts explored 22.3 miles (35.9 km) of the lunar surface at an average speed of 5 mph (8.1 km/hr). The longest mission sortie was 12.5 miles (20.1 km) reaching the greatest range from base of 4.7 miles (7.6 km). At the time, Schmitt and Cernan expected they would be the last humans on the Moon until the late 1980s. They left the Moon on December 14, 1972 carrying 253 pounds (115 kg) of samples, taking 2,120 photos, and having the distinction of being the last humans on the Moon.

THE APOLLO 15 THROUGH 17 COMMAND MODULE PILOT SPACESUIT AND DEEP SPACEWALKS

Besides the well-remembered lunar surface explorations, Apollo 15 through 17 also featured spacewalks on the return voyages by the Command Module Pilots. The Command Module Pilot was the only Apollo astronaut on every mission to the Moon who did not descend to the surface and walk on the Moon. Perhaps as a consolation, these astronauts on the last three missions were given a spacewalking task. This made these space pioneers the only humans to perform "deep-space spacewalks" in that they ventured into the vacuum of space (not being in low Earth orbit or on the Moon). Since this was a lesser objective of the Apollo 15 through 17 missions, simplicity and low cost were the key features of these EVAs.

As there was an abundance of Apollo 7 through 14-type lunar A7L suits and NASA was facing serious budget cuts, the decision was taken to modify earlier lunar suits for use by the Apollo 15 through 17 Command Module Pilots, thus producing the A7LB CMP configuration (Figure 9.26).

Except for four new-build pressure suits to support Apollo 17, all the Apollo 15 through 17 prime and backup Command Module Pilots used suits retrofitted from the existing Apollo 9 through 14 pressure suit inventory. The retrofit consisted of replacing the Multiple Water Connector (MWC) suit port with a plate-like plug (Figure 9.26), providing a new ITMG without an MWC hole, supplying A7LB gloves, and not supplying a Liquid Cooling Garment.

The new-build suits for Apollo 17 were identical to the retrofitted A7LB CMP except that the 400 series A7LB CMP suits lacked the plate filling the MWC hole.

For these spacewalks the primary life support system was provided by a 27.4-foot (8.3-m) umbilical attached to an inlet connector of the ILC suit and a Hamilton-made Pressure Control Valve attached to the suit's outlet connector. This simple single-feed hose with Pressure Control Valve provided a purge flow of oxygen with a nominal suit pressure of 3.7 psi (25.5 kPa). The purge gave ventilation flow through the suit, removed carbon dioxide and humidity, and provided cooling just like the Apollo backpacks that had been left on the

Figure 9.26. Command Module Pilot Suit without covers
(courtesy NASA)

Moon. The backup life support system for the lunar surface missions, the Oxygen Purge System, also provided a backup life support system for these deep-space missions (Figure 9.27).

During these deep-space tasks the Command Module Pilots were afforded the opportunity to leave the Command Module early on the return trip and traverse outside the spacecraft to the Scientific Instrument Module bay built into the side of the Service Module to retrieve film packages and experiments.

The first of these walks was supported by James Irwin who guided Command Module Pilot Alfred Worden's umbilical from the hatch of the Command Module on August 5, 1971. This allowed Worden to perform the world's first deep-space spacewalk some 171,000 miles (273,600 km) from Earth.

The Apollo 16 deep-space spacewalk was performed on April 25, 1972 by Command Module Pilot Thomas Mattingly. It lasted 1 hour and 24 minutes. The spacewalk permitted recovery of mapping and panoramic camera film packages, inspection of the spacecraft's exterior, and retrieval of a Microbial Ecological Evaluation Device. Before returning to the cabin, he opened his

Figure 9.27. The world's last deep-space spacewalk
(courtesy NASA)

visor briefly so he could see the stars, taking care not to look in the direction of the Sun.

During the Apollo 17 return voyage on December 17, 1972 Command Module Pilot Ronald Evans performed a 1-hour, 7-minute spacewalk to retrieve film from the Service Module. This had the dual distinction of being the last Apollo spacewalk and humankind's last deep-space EVA to date (Figure 9.27).

After the spacesuit items were delivered for Apollo 17, NASA funding for contractors dwindled to a trickle. Both Hamilton and ILC had to drastically reduce their workforces to about a quarter those at the peak of their Apollo employment. Hamilton's situation was perhaps slightly worse in that initially they did poorly in competing for Skylab business and accomplished limited recovery only in areas where the contract winners were unable to meet contract requirements. ILC fared little better. While NASA elected to use ILC Apollo pressure suits with minor modifications for Skylab and Apollo-Soyuz, the absolute minimum was ordered to support the missions. Apollo veterans from both companies caught up in the downsizing had to find employment elsewhere. Many did so in other parts of the country losing contact with former co-workers. One of the great Apollo ironies is that while ILC and Hamilton managements were unable to effectively work together without a NASA referee, personnel from the two companies worked side by side throughout the Apollo program with many becoming life-long friends.

By winning the Apollo Block II Pressure Suit and the Manned Orbiting Laboratory Suit competitions, the managements of Hamilton and ILC-Dover understood and respected the capabilities of each other's organizations. On the ILC side the New York-based International Latex management was no longer involved. Almost all the ILC-Dover management had risen from the technical ranks through their expertise at dealing with product issues.

In 1975 Hamilton Standard and ILC-Dover elected to voluntarily team together to compete for the Shuttle spacesuit contract. In 1977 they won. Most of the Apollo veterans of both organizations elected to return to become the technical core for the next American spacesuit that made all the Shuttle and American-suited International Space Station spacewalks possible.

CHAPTER 10

After the Lunar Missions

The Apollo suit did not end with astronauts returning from their last lunar visits of the 20th century. It has been said that the pressure suits used in the Skylab and Apollo-Soyuz were Apollo suits. Some credence can be given to this claim as there were only minor changes from Apollo configurations to better support the new mission roles. Perhaps another reason was that the Apollo-Soyuz mission was officially named Apollo 18.

THE SKYLAB SPACE STATION

After the United States reached the Moon the Soviet Union redefined the space race by channeling their efforts into being the first to have a manned orbiting space station. The United States took up the challenge. The Soviet Union's space station program was named Salyut. Salyut I was launched into space on April 19, 1971 but never reached orbit. Salyut II followed on April 3, 1973 but was damaged during launch. This station was never inhabited and was lost when it reentered Earth's atmosphere in fewer than two months. The U.S. space station was named Skylab (Figure 10.1). Skylab was also damaged during launch. Were it not for the fortunate selection of a spacesuit configuration that most NASA management thought was unnecessary, this space station would have been lost.

In 1969 some elements within NASA argued that program reliability was great enough to make spacesuits a waste of money. However, most believed that having a spacesuit to support launch and reentry emergencies was a minimum. The program did not plan extravehicular activities (EVAs). Thus, the debate centered on whether the program needed to be able to go out into space if there were unforeseen problems? As a result of limited budgets and time constraints, pressure suit selection was essentially being made by NASA's lunar program. Thus, using ILC-manufactured Apollo suits with minimal modifications for Skylab was essentially a forgone conclusion. However, which model would be the starting point: the lunar exploration Rover Suit or the Command Module Pilot's Suit? The answer hinged on the program having the ability to perform spacewalks.

Figure 10.1. Skylab Space Station concept drawing
(courtesy NASA)

While no one could prove EVAs were necessary, NASA fortunately decided to fund an emergency extravehicular capability. As a result, the Apollo 15 through 17 lunar exploration suit was selected for minimal modification (Figure 10.2). This reduced cost of certification to just the changes made.

Since Skylab spacewalks were just in case of an emergency, NASA planned an umbilical to provide life support. This was named the Astronaut Life Support Assembly (ALSA) and consisted of a chest-mounted Primary Control Unit, a leg-mounted Secondary Oxygen Package, and a 60-foot (18.3-m) umbilical assembly (Figure 10.3). The umbilical contained supply and return lines for cooling water, oxygen for pressurization and ventilation, electrical and communications lines, and a tether. AiResearch won the contract in November 1969. AiResearch, which became part of Allied Signal soon after the competition (now part of Honeywell), was the Command Module Environmental Control System provider. During Apollo, AiResearch entered into pressure suit development under internal funding to emerge the selected Apollo Block III spacesuit provider. This victory for AiResearch's Torrance, California Division was short lived as the Block III program was cancelled under mounting budget pressures. While small in terms of contract value the Skylab umbilical was a strategically important win for the people of Torrance.

To ferry crews to Skylab and return them to Earth, NASA used rockets and capsules originally purchased for Apollo. The space station was designed to be lifted into space by the Apollo program's Saturn V launch vehicle. The station was also called the Skylab cluster (Figure 10.1) as it was a series of modules

Figure 10.2. The Skylab Pressure Suit Assembly
(courtesy NASA)

Figure 10.3. The Astronaut Life Support Assembly
(courtesy NASA)

when in use. This consisted of the docked Command and Service Modules followed by the Multiple Docking Adapter, Airlock Module, Instrument Unit, and Orbital Workshop. Jutting out at right angles to the side of the Multiple Docking Adapter and Airlock Module was the Apollo Telescope Mount, which housed the solar observation cameras, and from which extended the four large solar arrays. The cluster had a total length of about 117 feet (35 meters) and its launch weight was around 199,750 pounds (90,606 kg).

The orbital paths of Skylab would sweep over an area that covered 75% of the Earth's surface, 80% of its food-producing regions, and 90% of its population.

The purpose of the Skylab 1 mission was to place the station in orbit. Skylab 2 was to bring the first crew a few days later. It and subsequent missions were to collect data on human habitation in zero gravity and conduct space science experiments. The science to be carried out aboard Skylab included biomedical and behavioral performance studies designed to evaluate human responses and capabilities in space under zero gravity at progressively longer durations. Other experiments and work tasks explored human-machine relationships, aimed at developing and accessing techniques for sensor operation, maintenance and repair, assembly and setup, and mobility required for longer duration space flights such as manned missions to Mars.

Experiments in solar astronomy, earth resources, science, technology, and applications rounded out Skylab's investigative agenda. As a by-product of operating for extended periods of time, information on how to increase the life of the various spacecraft operating systems would be gained.

Spacewalks were expected to figure prominently in Skylab's mission but only as a means of conducting experiments and retrieving data. The Skylab 1 launch of May 14, 1973 profoundly changed this expectation when the station was damaged about one minute after liftoff. Skylab 2 became a rescue and repair mission. If left unrepaired, Skylab would have been irreparably damaged by the buildup of heat in a matter of weeks.

Readings indicated the station was overheating and producing too little power to support manned use. The heat source afflicting the station was emanating from a section of the Orbital Work Shop. The likely cause was loss of the thermal outer covering in that area. Neither of the main solar arrays attached to the work shop were functioning.

The solution was a parasol devise that could be deployed through an existing portal in the affected area of the work shop. ILC, the Skylab pressure suit provider, immediately sent two engineers, one fabricator, and all the needed materials to Huntsville, Alabama to develop and fabricate a thermal repair solution. The fabricator upon whose shoulders this daunting task fell were those of Ellie Foraker. Huntsville was essential to the development because it had a full-scale training replica of the station that allowed testing the parasol before it launched. In just nine days, this emergency thermal protection system was taken through prototyping to production of the flight unit with training and flight spares plus the accompanying tools. All of which were designed, made, tested, used by the crew for training, packed for shipment to Cape Canaveral, shipped, unpackaged, and stowed for launch in the allotted time.

Eleven days after Skylab was launched the Skylab 2 rescue mission lifted off in what was an historic experiment in space repair. Once in orbit the Skylab 2 crew maneuvered their spacecraft close to the injured station to assess the damage. The astronauts reported that Solar Array Wing 2 (Figure 10.4, upper left) and most of the meteoroid shield over one side of the Orbital Work Shop (Figure 10.4, center) were gone. Solar Array Wing 1 appeared intact. However, it was in the folded position being held down by a metal strap (Figure 10.4, lower right).

The Skylab 2 spacecraft was then carefully guided so close to the jammed solar array on the station that Astronaut Paul Weitz could stand with his upper body through a hatch and attempt to free the array with a 15-foot (4.5-m) pole with a "shepherd's hook" on the end. Weitz pulled on the array while Astronaut and Medical Doctor Joseph Kerwin gripped his legs. All the time, Astronaut Charles Conrad attempted to keep the spacecraft in position as Weitz's efforts were pulling the craft toward the station. When that proved unsuccessful, Weitz replaced the hook with a universal prying tool. The strap still did not budge. Thwarted again the astronauts docked their craft with the station to rest since they had been working for 22 hours.

Once inside, the crew deployed the parasol sunshade by pushing it out through the scientific airlock in the side of the work shop and unfolding it. Once it was in place, the interior of the work shop quickly began to cool to a

Figure 10.4. Solar array and thermal shielding missing
(courtesy NASA)

tolerable 80°F (26.7°C). However, to make the station operational, more electrical power was needed.

In parallel with the activities of the first day and sleep period in orbit, a ground-based team of astronauts and engineers devised, assembled, and tested a strap-cutting devise made up of tools and equipment that were available in orbit. The ground-based team then transmitted the assembly instructions for the devise to the astronauts in orbit while they were sleeping. The devise was a cable cutter mounted at the end of a 29.5-foot (9-m) rod and actuated by a rope. Ground control personnel surmised that another problem facing the astronauts was that a hydraulic damper was probably frozen in position. Considerable force would have to be exerted by the crew to free it. The technique developed by the ground-based team consisted in the crewmembers pushing up against the mid-section of a rope strung between the solar array and an attachment point on the station.

Freeing the jammed solar array was the first priority of the next morning and Astronauts Conrad and Kerwin soon exited the station for the repair mission. In space, things did not work quite the same as in the ground simulations. The cable cutters were used to clamp onto the strap, and Astronaut Conrad then translated along the rod to cut the strap with a surgical bonesaw. To free the frozen actuator, both crewmembers strained upward against the

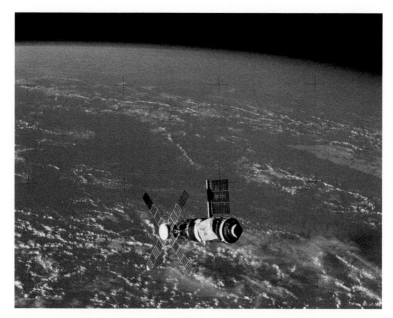

Figure 10.5. Skylab after the first spacewalk
(courtesy NASA)

rope and freed the beam. In Conrad's words, when the beam let go, both astronauts "took off." The newly freed array started generating power immediately,
which was indicated by the ground's terse but happy announcement: "SAS
amps!"

In addition to freeing the Solar Array Shield beam the astronauts changed
out a film magazine for the Extreme Ultraviolet Coronal Spectroheliograph
and pinned open the door for the X-Ray Spectrographic Telescope. The only
extravehicular problem worthy of note was some difficulty experienced by the
astronauts in restowing their umbilical life support systems in the stowage
spheres. The spacewalk lasted some 3 hours and 25 minutes and resulted in a
fully functional, soon-to-be-operational space station (Figure 10.5).

On June 19, 1973 the second and last spacewalk of Skylab 2 was conducted
by astronauts Weitz and Conrad. This lasted 1 hour and 36 minutes. The crew
returned to Earth after 28 days in space, which set a very short-lived world
record for the longest time spent by humans in space.

Skylab 3 was launched on July 28, 1973. On August 6, 1973 Astronauts
Garriott and Lousma (Figure 10.6) undertook a 6-hour, 31-minute EVA (SL-3/
EVA 1) to install the thermal shade and service a number of experiments.

The shade was installed over two poles, assembled in orbit out of 5-foot
sections mounted in a "V" shape, and extended some 55 feet (16.8 m) over the
Orbital Work Shop exterior. The spacewalkers then progressed to exterior
inspections and experiments. This EVA, which lasted 6 hours and 31 minutes,
left the space station in its final form (Figure 10.7).

Figure 10.6. Lousma in first Skylab 3 spacewalk
(courtesy NASA)

Figure 10.7. Skylab after Skylab 3
(courtesy NASA)

Skylab 3 continued its exploration mission with a test undertaken inside the space station of a propulsion backpack called the Automatically Stabilized Maneuvering Unit. This was followed by another spacewalk for maintenance and experiments and another test inside the station of a Foot Controlled Maneuvering Unit. The tests of the maneuvering units were conducted inside the Orbital Work Shop. The astronauts of SL-3 spent 59 days aboard Skylab before their return on September 25, 1973 setting another short-lived world record. This achievement also set the stage for the Skylab 4 science-focused mission that extended the record yet again to 84 days and included many zero-gravity experiments.

A major part of the Skylab story was the human ability to overcome mistakes. Skylab faced never-ending challenges after its launch but still made significant technical contributions. Spending 41 hours and 46 minutes outside the station the two-person crews of Skylab not only fulfilled the original plans of servicing and conducting experiments, but also performed repair and outfitting tasks that guaranteed the program its very existence. As one astronaut said: "Skylab worked better broken than anybody had hoped for if it was perfect." This was effectively demonstrated by the fact that the last crew, although they had flown by far the longest mission, returned to Earth in the best physical condition.

For those who doubted whether the cost was justified, it is worth noting that just one copper deposit in Nevada located by Skylab was estimated to have an ultimate value of billions of dollars, far exceeding the cost of the whole U.S. space program up to that point.

In 1974 Skylab was closed. However, this was to be temporary. NASA's plan was to make the next spacecraft, the Space Shuttle, operational while Skylab remained in orbit. The Space Shuttle was to rescue Skylab from reentering Earth's atmosphere and support its continued use. The rescue plans were based on a booster rocket system named the Teleoperator Retrieval System that was to be carried into orbit by the Shuttle and attached to Skylab by the Shuttle's robotic arm. The Teleoperator Retrieval System was to boost the station into higher orbit for later manned use. Some members of Congress thought that NASA was predicting an unrealistically early orbital decay and reentry of Skylab to boost its annual Shuttle program budgets and found their own experts who predicted a longer orbital survival for Skylab. Consequently, Shuttle development was pushed out. The Teleoperator rocket engine module and Shuttle were designed and built, but Skylab did not wait for them to be mission ready. On July 11, 1979 the world watched as Skylab became a fireball that hurtled toward Earth scattering debris over the Indian Ocean and Australia.

ENDING APOLLO IN SPACE COOPERATION, APOLLO-SOYUZ

Apollo started as a non-military, alternative competition between the U.S and the Soviet Union who appeared to be on the brink of war at the beginning of the 1960s. In the race to the Moon the Soviets were the first to place a man in space and in orbit around the Earth. They were also the first to orbit two people and rendezvous two spacecraft. Despite these milestones the U.S was close behind though always trailing. This race ultimately took a human toll on both sides. In 1961 Cosmonaut Valentin Bondarenko died in a ground test fire under sea-level pressure and a pure oxygen atmosphere. The Soviet government made safety improvements in their space program and successfully concealed the loss from the world.

By 1966 the space race hinged on which side would be first to develop a heavy-lift, lunar rocket. The U.S. appeared closer to reaching that milestone first. Unfortunately, in January 1967 Chaffee, White, and Grissom died in a test fire under conditions similar to the 1961 Soviet disaster. Subsequent Soviet propaganda touted their technical superiority and spoke of U.S. incompetence. The U.S. completed their safety reviews and became operational with the first Apollo manned flight in October 1968.

It was also in that month that Dr. Thomas Paine succeeded James Webb as the Director of NASA. Paine was an advocate for profound change in relations between the U.S. and Soviets, specifically cooperation in space. In April 1969 Paine began official correspondence with Anatoliy Blagonravov, the Chairman of the Soviet Academy of Sciences. This started a dialog about space cooperation that was not fully endorsed by either of their governments.

The Soviets were unable to successfully develop their heavy-capacity lunar rocket before the U.S. reached the Moon. To progress their program in advance of completing their lunar rocket, the Soviets essentially took an existing spacecraft designed for two and were able to launch three cosmonauts in January 1969 by removing "unneeded" mission-specific systems.

The Soviet Space Agency deemed pressure suits for launch and reentry unnecessary for Soyuz 11, which launched in June 1971. On June 30, 1971 the spacecraft lost internal cabin pressure during the reentry process. Lacking even the most basic of spacesuits Cosmonauts Georgy Dobrovolsky, Vladislav Volkov, and Viktor Patsayev perished. Director Paine's campaigning bore fruit; U.S. President Nixon sent a conciliatory message to the Soviet people. Astronaut Tom Stafford was sent to be one of the pallbearers at the cosmonaut's funeral. With these gestures the Apollo-Soyuz Test Project gained the political acceptance needed to become a reality.

Apollo-Soyuz had practical, technical motives for linking up the two predominant spacefaring nations in the first international manned space flight. Developing the ability for rendezvous and docking between American and Soviet spacecraft could allow rescue by a craft of the other nation's crew in danger or distress. There were formidable challenges facing this endeavor. Aside from the politically adversarial positions of the two nations, there were language and geographical barriers, plus radically different approaches to technical solutions to human space exploration. While this rescue capacity was not subsequently used, the processes that were developed for addressing the safety and reliability issues between these two dynamically different space agencies previewed the path taken for the joint development of International Space Station systems over a quarter century later.

Since the mission name of the Apollo-Soyuz Test Project was Apollo 18, the

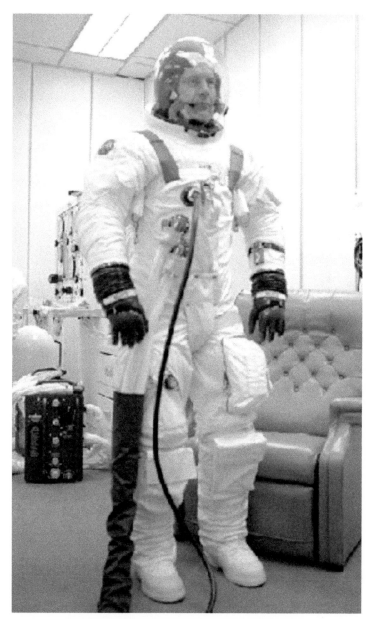

Figure 10.8. Deke Slayton ready for launch
(courtesy NASA)

program's Pressure Suit Assembly (Figure 10.8) was indeed an Apollo suit, although that is typically missed in most discussions.

The preceding Apollo configuration selected to provide the basis for the Apollo 18 suits was the Command Module Pilot suit used on Apollo 15 through 17. Since there was to be no spacewalk, the suits for the Apollo 18 crew of Thomas P. Stafford, Vance D. Brand, and Donald K. "Deke" Slayton were modified to reduce weight and cost. The normal outer cover layer of Teflon Beta cloth with underlayers of aluminized Kapton with nylon spacers was replaced with Teflon Beta Polybenzimidazole fabric, which increased its durability. Moreover, extravehicular gloves, the positive pressure relief valve, the extravehicular visor, and the connectors for the Liquid Cooling Garment and Emergency Oxygen System were removed.

The single Test Project mission launched on July 15, 1975 and marked the successful debut of the first international cooperation in manned spaceflight through the linkup of Russian and American vehicles.

The Apollo-Soyuz Test Project verified the compatibility of rendezvous and docking systems for American and Soviet spacecraft. It not only presented the possibility for international space rescue during a key period of the Cold War, but ushered in the possibility of cooperation and peaceful coexistence that repeatedly emerged in the decades that followed.

WHERE DID THE "CREATORS" OF THE APOLLO SPACESUIT GO?

Some of the people who contributed to the spacesuit first used on the Moon were not part of the final program. As a result of NASA electing to save money by upgrading existing pressure suits and backpacks, the later Apollo and Skylab missions meant less employment for the Apollo spacesuit creators. The funding lapse between Apollo and Skylab resulted in most of the remaining spacesuit people becoming unemployed. However, these were creative, hard-working, and talented people who adapted to changing times and got on with their lives. Perhaps the most intriguing part of the Apollo spacesuit story is: Where did they go?

Iona Allen had been key to production of the Apollo Thermal Outer Garment, but at the end of Apollo, ILC's Dover operations dwindled from 600 to a staff of just 35. Allen was one of the hundreds who lost employment. However, she elected to return to ILC when the business improved years later and was able to use her talents to help create the Shuttle Extravehicular Mobility Space Suit (a.k.a. the "spacewalking" spacesuit) before retiring.

Earl R. Bahl and the space-related workers at Hamilton Standard did better than ILC. Of the some 750 that supported space programs, Bahl and 500 colleagues remained at Hamilton Standard after the end of Apollo. When space employment rebounded in the late 1970s, almost all Hamilton's laid-off workers returned to space work. Bahl went on to support many other programs including the Shuttle Extravehicular Space Suit. He retired as the head of Space System's Mechanical Design group. In spite of the long hours he spent on space programs, Bahl is still happily married to his Apollo bride. He was a reviewer of this book.

Ronald J. "Ron" Bessette lost his employment with ILC as part of Apollo downsizing, but returned many times as a contractor to leave his fingerprints on many space items before retiring. Bessette still loves sailing, was a reviewer of this book, and was invaluable in restoring the record of contributions made by ILC's people.

Bradford "Brad" Booker was fortunate to have post-Apollo continuous employment supporting many space programs. Booker retired from Hamilton Standard as a mechanical design engineer.

Evelyn "Ev" Case was known as Ev Kibbler during Apollo. After Apollo, Ev divorced and married an ILC engineer named Mel Case. She remained at ILC and was the pattern maker through Shuttle. Engineers mentored by Ev serve ILC today.

Melvin C. "Mel" Case continued to be a respected member of the ILC design team making additional contributions to Shuttle spacesuits before retiring as a result of health issues.

Russell S. Colley was recruited by NASA to lead Mercury spacesuit development. Colley retired from NASA and returned to Ohio. In retirement, he practiced jewelry design and became a respected watercolor artist. Colley received 65 patents during his career and NASA's Distinguished Public Service Medal in 1994. He died February 4, 1996, in Springfield, Ohio.

George P. Durney retired from ILC and lived to see some of his work used in building the International Space Station. In recognition of his contributions, ILC annually bestows the George P. Durney Pride Award to the year's outstanding employee. His son elected to join ILC and remains part of the organization.

John Flagg succeeded David Clark as President of the David Clark Company. He retired from David Clark in 1992 after overseeing the development and production of the Space Shuttle launch and reentry spacesuits.

Eleanor "Ellie" Foraker supported ILC-Dover for 43 years before retiring. With the Shuttle era, Ellie had the opportunity to extend her mentoring to female engineers who went onto become Senior Engineers on the Shuttle and International Space Station programs. Ellie also left her fingerprints on the Mars Pathfinder program. With retirement, she spent the remainder of her life dedicated to her family, especially to her grandchild and great-grandchildren.

Douglas E. "Doug" Getchell lost his employment with Hamilton as part of Apollo downsizing. Former co-workers eventually lost contact. He has not been successfully relocated in the creation of this book.

Jerry R. Goodman moved on to other NASA efforts during Apollo and continued with NASA in a cascading series of programs and assignments until he retired November 2015. Goodman is now a NASA Emeritus reporting to the Space & Life Sciences Division Chief. He was a reviewer of this book.

Andrew J. Hoffman made the move to Hamilton's Lunar Module effort where he became the Program Manager. This was after Hamilton Standard lost the pressure suit and integration portions of the Apollo spacesuit. He retired from Hamilton as Executive Vice President, after which he started an engineering services company thanks to the understanding and support of his Apollo bride. He volunteered the author to be an unpaid space historian in 1992. He is now entirely retired, happily married, and was a reviewer of this book.

David C. Jennings worked into the Shuttle era before retiring from Hamilton Standard. He and his wife Eleanor moved back to his native New Hampshire where he lives today. He was a reviewer of this book.

Eleanor J. "Lee" Jennings graced humanity with her presence for 87 years, 65 of which were married to David. A nurse by education, she was never a directly paid member of the spacesuit community, but her contribution illustrated that the Apollo program was only possible through family support.

Dr. Robert L. "Bob" Jones had a Ph.D. degree, which was not common during Apollo. Former colleagues believe he left NASA to pursue academic endeavors.

J. "Al" Kenneway was caught up in the post-Apollo downsizing at ILC. Al returned to Massachusetts and became lost to American spacesuit history.

Ceil Kenneway became Ceil Webb (*see* **Ceil Webb**).

Evelyn "Ev" Kibbler came to be known as Ev Case (*see* **Evelyn "Ev" Case**).

Beverly "Bev" Killen was a key member of staff at ILC, so much so that she remained employed during the post-Apollo downsizing. Her career continued into the Shuttle spacesuit and many other projects before retiring from ILC. She was loved by everyone.

John J. Korabowski was a non-degreed engineer. He survived the post-Apollo downsizing by taking a technician position in one of Hamilton Standard's labs. He quickly rose to be a Space Lab leadman, where he remained until he retired in the Shuttle era. Korabowski left records with family members that have been incorporated in this book.

Stanley "Stan" Krupinski was a key contributor to development at B. Welson & Company. The company was wound up before the end of the Apollo program, thus scattering its workforce. We do not know where he went next. However, his method of attaching cooling tubes to Apollo Liquid Cooling Garments was transferred to Delaware in 1969 and is still used in American spacesuit manufacture today.

Carroll Krupp attended Akron University and was a man of many parts. Besides his contributions to U.S. pressure suit development, he was well known for flying model airplanes and building his own sports car. He accrued over 50 patents during his career. When he retired from B.F. Goodrich after 46 years of service, he spent his time on woodworking projects for his wife, six daughters, and his church. Krupp was a devoted family man who had 14 grandchildren and 4 great-grandchildren at the time he passed away at the age of 87.

John S. Lovell went on to head an Advanced Engineering Group that supported space and defense activities. Retiring from Hamilton in 1986, he continued in contract engineering for a number of years before finally retiring. Lovell supported a Japanese documentary on the development of the Apollo spacesuit in 2012 and made a contribution to this book.

Michael A. "Mike" Marroni Jr. was one of the "suit people" for whom Hamilton Standard tried to find alternative internal employment when the Manned Orbiting Laboratory program was cancelled in 1969. However, he found the new work less interesting and left for a career in nuclear power generation. When last contacted, he was trying to spend as much time golfing as possible.

James W. "Jim" McBarron II was heavily involved in the Apollo program at NASA, especially with Skylab, Shuttle, and the International Space Station. He left in 1999 to become a spacesuit systems manager for advanced spacesuit technology concepts at ILC-Dover. He then left ILC in 2002 to start his own consulting service to support development of advanced spacesuit technology and inflatable products for current and future manned space missions.

James H. "Jim" O'Kane and his wife moved to Grand Cayman Island on retirement from NASA. At last contact, he was parking their recreational vehicle (small self-propelled home) in Houston. When the weather and mood took them, they would hop on a plane from Grand Cayman and fly to Houston to recommence touring. When they were ready once more for the Caribbean, they parked their recreational vehicle in Houston and returned home.

Rico Perry was not one of the lucky 35 ILC people who retained employment after Apollo. She did not remain in contact with other ILC veterans, thus we do not know where she went next.

Roberta "Bert" Pilkenton was a production sewer at ILC. She came back to work when Shuttle funding started and played a major role in the creation and production of the Shuttle Extravehicular Mobility Unit (a.k.a. spacesuit).

Anna Lee Pleasanton bought two surplus industrial sewing machines from the idled Dover plant and started a drapery business. When ILC business picked up, she elected not to return to spacesuits but continued with her shop in Milford, Delaware.

Richard C. Pulling and **Mel Case** started an aerial photography business on the side during Apollo. Case grew tired of the business, but Pulling pressed on alone. The post-Apollo layoff gave more time for the business. One day, Pulling was in tourist shop looking at postcards. He noticed that most cards depicted things that no longer existed or showed automobiles that clearly dated the pictures as extremely old. He got the idea to expand the photography business into printed cards. Within a year he was spending his weekends restocking stores. He got his wife and two sons involved, and then the family business expanded to include other products. Pulling never returned to spacesuits.

George C. Rannenberg III continued at Hamilton to accrue 32 patents before leaving the firm. Not ready to give up engineering, George spent five years as a consulting engineer in Italy working on European Space Agency projects before finally retiring. He is well remembered.

Dixie Rinehart relocated to the Rocky Mountains following the Apollo downsizing. Now retired, Dixie enjoys going to vintage car meets to check out the Lincolns on display.

Donald "Don" Robbins was principally responsible for David Clark Company spacesuit soft-goods design in the mid-1960s. He left the David Clark Company in the late 1960s to start a business of his own.

Joseph "Joe" Ruseckas succeeded John Flag as the Director of Research and Development in the early 1960s and retired in 1986.

Richard "Dick" Sears joined the David Clark Company in the late 1950s to assist with the company's increasing pressure suit business. He quickly grasped the nuances of both partial and full-pressure suits, making valuable contributions to numerous suit programs during his time at the David Clark Company. He left in the early 1970s to collaborate with David Clark Associates. He was David M. Clark's son-in-law.

Lenard F. "Len" Shepard went on to get involved in a seemingly endless series of projects that caught his fancy. He never returned to spacesuit work.

David H. "Dave" Slack loved his wife, his family, being a spacesuit engineer (he kept going back when there was funding), archery (he was a nationally ranked competitor), and driving a school bus for his wife's company when there was no suit work (he loved kids). He was a good friend to many and is sorely missed.

Elton Tucker joined NASA to become Head of the Suit Section for the Gemini Support Office. After Gemini, he left NASA to join a metal bellows manufacturer on the east coast.

Ceil Webb remained with ILC after Apollo, got divorced from Al Kenneway, and remarried. At the end of a distinguished ILC career, she moved to Florida to enjoy the sun in her retirement.

Ruth Ann Wilkerson retained employment with ILC during the dark days following Apollo and went on to help develop the Shuttle spacesuit and many space and industrial products. At the time of writing this book she was still working at ILC.

Hilda Willis went back to bar tending after Apollo and elected not to return to spacesuits.

Robert C. Wise continued at Frederica, Delaware to be part of the 35 who were retained between the Apollo and Shuttle programs. Upon retirement from ILC, he retired to Ohio.

INFLUENCES ON FUTURE HUMAN SPACE EXPLORATION

The space race between the two greatest powers of the 20th century started when President Kennedy determined to have a human set foot on the Moon before 1970. When the challenge was thrown down, humankind did not know how people could survive while walking on the Moon, let alone effectively perform scientific studies. The creation of the Apollo spacesuit was a long, hard struggle. Every step was a test of human ability to overcome obstacles by innovation, determination, and perseverance. Hundreds of people from many organizations made thousands of contributions. Some were significant, but this massive achievement of the 20th century which brought the space race to a wondrous conclusion would not have been possible without all those efforts.

The Apollo spacesuit had greater mobility than any preceding space pressure suit. It was the first U.S. spacesuit system to use an autonomous life support system for extravehicular activity. Apollo spacesuits reliably and successfully supported Apollo missions in every possible way. The Apollo 7 through 14 configurations made the first human explorations of the Moon possible. The Apollo 15 through 17 spacesuit improvements facilitated the most extensive, first-hand lunar surface study in the history of mankind.

The legacy of the Apollo spacesuits is that they made historic events possible, they influenced the human space exploration that followed, and they will continue to influence such explorations for decades to come.

Despite receiving little public attention at the time, the significance of astronauts falling over on the lunar surface with the potential of puncturing the suit's exposed life support system hoses was not lost on NASA management. When James V. Correale became Chief of Johnson Space Center's Crew and Thermal Systems Division, he vowed that never again would the conduits carrying ventilating gases between the suit and its associated life support system be exposed. This vow came true in the next new-suit approach: the Space Shuttle's

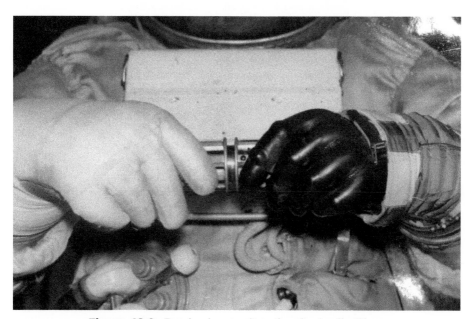

Figure 10.9. Russian lunar suit testing the Apollo Glove
(courtesy Nik Moiseev)

Figure 10.10. Russian Sokol Model KV-2 "Rescue Suits"
(courtesy Gary L. Harris)

extravehicular spacesuit featured a direct connection between the pressure suit and the life support system backpack without hoses.

In many ways the efforts made in creating the Apollo spacesuit have been the point of reference for all subsequent American spacesuit activities. However, that influence was not confined to the United States.

During the Apollo-Soyuz Test Project many friendships were formed between astronauts and cosmonauts. In 1972 Astronaut Michael Collins pre-

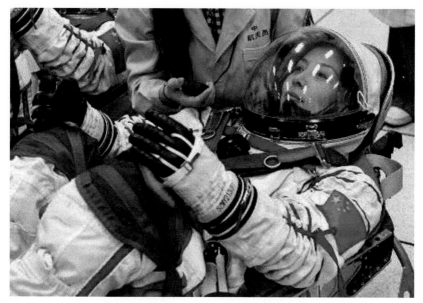

Figure 10.11. Taikonaut Wang Yaping
(courtesy China National Space Agency)

sented an obsolete, Dixie Rinehart-designed, ILC Apollo glove to Cosmonaut Vitaly Sevastyanov as a gift. Once home in Russia, Sevastyanov could not resist seeing whether the glove could actually hold pressure. When it did, Sevastyanov had to see how it performed under test conditions. The Apollo glove was found to be more mobile and flexible than the Soviet gloves of the time.

Sevastyanov then showed the glove to Guy Severin, General Designer of the Soviet Union's spacesuit factory, with the words "The American glove is better than yours." Severin asked to borrow the glove to see for himself. Sevastyanov loaned him it, but never got it back.

The factory, now NPP Zvezda, made an interface to connect the glove to the arm assembly of a Soviet lunar spacesuit, their alternative to the American Apollo suit (Figure 10.9). Testing under pressure showed that Sevastyanov was correct. Severin decided to disassemble the glove. All parts were carefully copied. Severin then married the Apollo glove and wrist to a Soviet glove disconnect, as he judged his disconnect to be better. The result was a pressure glove that first saw service in the Soviet Orlan spacesuit and the early spacewalks off the Salyut space stations. This escaped the attention of the U.S. as it was hidden from view by the Soviet-designed outer thermal glove.

These Soviet Apollo replicas reached space service with the first Salyut spacewalk on December 20, 1977. They supported the first woman to perform extravehicular work, Svetlana Savitskaya, on July 25, 1984 outside Salyut 7. This design continued in use on the Mir space station until it was replaced by a

Soviet-designed glove in 1988. However, this was not the only Soviet spacesuit to be influenced by the Apollo suit.

The Soviet factory also used the glove with a slightly simplified wrist on the Soviet Sokol Model KV-2 Rescue Suit prototypes starting in 1973. The Apollo-based glove design continued to be used on the Sokol KV-2 (Figure 10.10) until 2003-04 when an all-new Zvezda-designed pressure glove was introduced. However, this did not represent the end of the Apollo glove trail.

Yet another offshoot of the Apollo glove story came in the early 1990s when China bought two Russian Sokol KV-2 suits to support taikonaut (the Chinese equivalent of astronaut and cosmonaut) space program training. The Chinese then reverse-engineered the Russian suits to produce the spacesuit used on Shenzhou V, China's first manned space flight. While the torso assemblies have undergone changes due to subsequent Chinese developments the Apollo-based glove design is still in service (Figure 10.11). Perhaps the ultimate irony of this international cross-pollination is that the key goal of the Shenzhou program is that the next humans to set foot on the Moon will likely be Chinese.

APPENDIX I

The Origins of the Apollo 11 Spacesuit Technologies

No one person or organization invented the Apollo spacesuit formally called the Extravehicular Mobility Unit or "EMU." The inspirations that overcame countless technical challenges originated from many sources over many years. While the public is generally familiar with the exterior features of the Apollo spacesuit shown in Figure A1, most of the inventions and innovations involved in its construction are visible only when the exterior garments are removed (see Figures A2 and A3). Reference numbers 1 to 25 identify spacesuit features, and the corresponding notes identify individuals and organizations that made those contributions to the first spacesuits used on the Moon.

1. *Pressure Suit Assembly*—The suit was developed and manufactured by International Latex Corporation (ILC) in Dover, Delaware. In 1968 ILC initiated a patent application that NASA subsequently approved (U.S. Patent 3,751,727). While it reflected the ongoing rivalry with Hamilton Standard, it also identified ILC employees who were key to development of the Apollo pressure suit. These people, as appears on the patent, were Lenard F. Shepard, George P. Durney, Melvin C. Case, A. J. Kenneway, Robert C. Wise, Dixie Rinehart, Ronald J. Bessette, and Richard C. Pulling. In 1969 International Latex partially sold its pressure suit organization to form what is now ILC-Dover LP. ILC-Dover relocated to Frederica, Delaware in two stages starting in 1969. International Latex sold its remaining interest in ILC-Dover in 1984. Reference numbers 3–6 and 12–25 explain features embodied in the Apollo pressure suit.
2. *Portable Life Support Assembly*—The backpack was a separately donnable life support system assembly that was developed and manufactured by the Space and Life Systems unit of the Hamilton Standard Division of United Aircraft located in Windsor Locks, Connecticut. As a result of the acquisitions and mergers of Sundstrand and Goodrich Corporations in 1999 and 2012, respectively, the Division became Hamilton Sundstrand and then United Technologies Corporation (UTC) Aerospace Systems. The remnants of the unit that supported Apollo still exist in Windsor Locks. Reference numbers

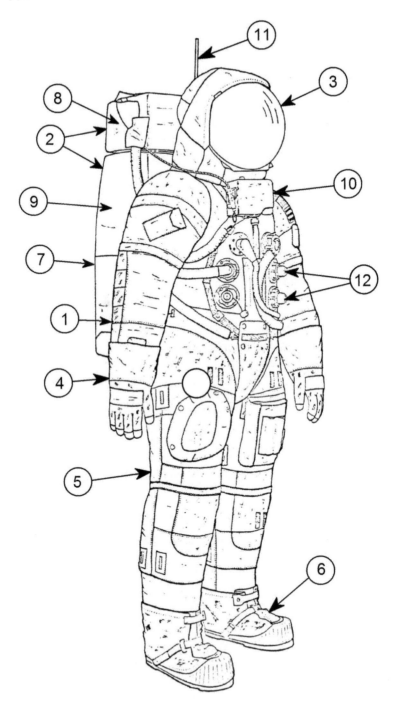

Figure A1. Apollo spacesuit with cover garments
(courtesy NASA/K. Thomas)

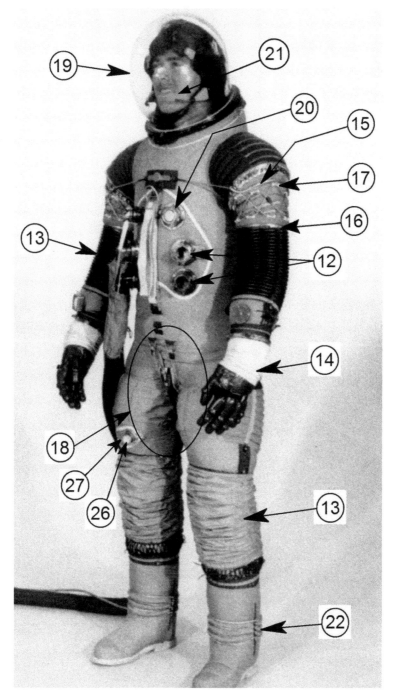

Figure A2. The ILC A7L pressure suit without cover garments
(courtesy NASA/K. Thomas)

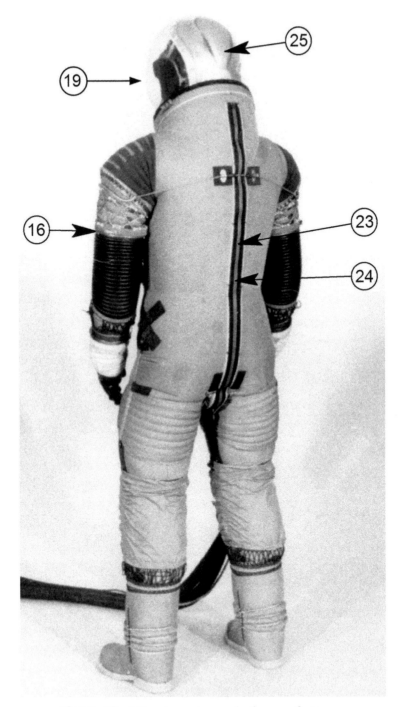

Figure A3. A7L entry system and other rear features
(courtesy NASA/K. Thomas)

7–11 explain the subassemblies and features embodied in the Apollo backpack.

3. *Lunar Extravehicular Visor Assembly (LEVA)*—The gold-coating process for the eye-protecting Apollo sun visor was developed by Perkin Elmer Corporation in Norwalk, Connecticut starting in 1963 under contract to Hamilton Standard. ILC initially was responsible for the remainder of the visor assembly design and development. Hamilton assumed visor design and prototyping in November 1963. Analysis at Hamilton indicated that thermal insulation would be required over the entire helmet area. This was initially planned to be provided by the hood of an outer-garment that resembled a parka. A fiberglass outer shell was incorporated to support the weight of the garment. This was revised in 1965 to having a thermal cover directly attached to the outer shell of the visor assembly. The resulting visor cover extended down to cover and wrap around the neck ring. In parallel, NASA engineers O'Kane and Jones designed their own helmet with a LEVA. Their analysis indicated the polycarbonate layers of the LEVA and helmet would provide sufficient thermal insulation that thermal outer covering was unnecessary. They selected Air-Lock Corporation in Milford, Connecticut as the manufacturer of their prototype visors and Apollo program visors after 1965. Thermal transfer issues caused this design to gain an ILC-provided neck ring thermal cover by 1967. As a result of condensation in the helmet during Apollo 9 shaded periods, the visor assembly was revised to have a fiberglass outer shell, a 1965-style thermal cover, and side opaque sun shields that could retract or be flipped down as needed. This configuration was used for Apollo 11 through 13. Apollo 14 through 17 used a derivation that added an opaque center visor.

4. *Integrated Thermal Outer Gloves*—The David Clark Company was the advocate for integrated glove coverings rather than separately donnable outer gloves. At NASA's request, the David Clark Company educated Hamilton and ILC on its outer glove construction methods for Apollo. The techniques for layering and securing the insulation material in the outer glove were developed by NASA and the David Clark Company for the Gemini program. The David Clark Company transfered these developments to ILC at NASA's request. The tactile, reinforced, molded RTV (room temperature vulcanite, a.k.a. "rubber") fingertips were an ILC innovation and development.

5. *Torso Thermal Meteoroid Garment*—The David Clark Company was the advocate for torso coverings being an integrated part of the pressure suit rather than separately donnable outer garments. ILC adopted this system for Apollo in 1967.

6. *Extra Vehicular Over Boots*—The Apollo 11 through 17 style of thermal overboot was introduced by ILC. The overboot used Chromel-R, a stainless steel wire fabric, on the outer surfaces and thermal insulation techniques introduced by the David Clark Company during Gemini that were substantially refined by ILC. The reinforced, molded RTV soles were an ILC innovation and development.

7. *Primary Life Support System*—Development included invention of the Porous Plate Sublimator (U.S. Patent 3,170,303, John S. Lovell and George C. Rannenberg inventors), which allows rejection of body and life support system heat to space with no moving parts. This system also supplied oxygen, removed carbon dioxide, controlled humidity, and cooled both ventilation gas and water for Liquid Cooling Garments. It was developed and manufactured by Hamilton Standard.

8. *Oxygen Purge System*—Nominally, a 30-minute backup life support system of the Apollo spacesuit. It was developed and manufactured by Hamilton Standard.

9. *Backpack Thermal Outer Cover*—The basic construction techniques of these outer covers were provided by the David Clark Company. The final material selections were made by NASA. These covers were manufactured by Hamilton Standard.

10. *Remote Control Unit*—The chest-mounted display and control system for the backpack life support system. It was developed and manufactured by Hamilton Standard.

11. *Antenna*—Radio antenna breakage was an early program challenge and of great concern as it would result in loss of or greatly reduced communications. The durability and flexibility of a Stanley Tool Company tape measure led to the purchase and space certification of a roll of tape measure raw stock to support the communications of men on the Moon.

12. *Gemini Style Life Support Connectors*—The Apollo program began considering these connectors in 1964. The Air-Lock Corporation developed the Gemini connector for the David Clark Company. NASA and ILC elected to make them the Apollo program connector in late 1965.

13. *Molded Convolute Joint*—The base design of this molded convolute originated with B.F. Goodrich as Russell Colley's "Tomato Worm" mobility element in the early 1940s. International Latex Corporation recruited Goodrich personnel in the early 1950s, by which time Goodrich was evolving past molded convolutes in favor of fabric mobility joints. ILC's George Durney, who was not a former Goodrich employee, refined and continued development of this type of molded convolute to make it a lower effort mobility system (U.S. Patent 3,432,860).

14. *Molded Pressure Gloves*—From World War II to the early 1950s, B.F. Goodrich made pressure gloves in single-layer, molded rubber with a fabric mesh substrate. In 1962 ILC competed for Apollo with a separate bladder and restraint layer (leather) glove system. For the Apollo program, ILC developed a more advanced single-layer, molded rubber glove with fabric mesh substrate. Palm bars, which allow the palm area to bend for grasping tools, were initially developed in the World War II MX-115 program and used in Goodrich Mercury gloves. The first Apollo (1962–63) multidirectional wrist joint (by ILC) was a single-cable arrangement. The second Apollo multidirectional wrist joint was created by Hamilton Standard in 1964 featuring a redundant Teflon cord and ferrule wrist design. The remainder of the Apollo glove was a

copy of the preceding ILC Apollo design. For the 1965 competition, ILC developed a third Apollo design that had a redundant, two-cable, multi-directional wrist joint, which became the Apollo program glove when ILC won the 1965 competition. In these and later ILC gloves, Teflon sleeving in the metallic wrist restraint conduits reduced the effort needed to move. Teflon sleeving of metallic restraint conduits appears to have been a Hamilton innovation that was adopted by ILC.

15. *Shoulder Cable Restraint System*—This was probably first seen in the U.S. MX-115 pressure suit development program of World War II. It enjoyed popularity with Goodyear and Republic Aviation. Early attempts with this approach experienced excessive force to achieve mobility. ILC experienced similar problems in their first four Apollo suit designs. Teflon sleeving in the metallic restraint conduits reduced friction allowing acceptable mobility. This appears to have been a Hamilton innovation that was adopted by ILC.

16. Pressure (lip) Sealed Upper Arm Bearings—These were features in the B.F. Goodrich 1953 Omni-Environment Inflatable Suit (U.S. Patent 2,966,155, Carroll Krupp inventor) and in the Goodrich Mk. II suits that followed. Unfortunately, the cramped cockpits of fighter aircraft made hard contact with these bearings unacceptable. Goodrich brought upper-arm bearings to the Apollo program in late September or early October 1964 in a joint Goodrich/Hamilton Standard prototype effort. The initial negative NASA reaction delayed consideration of the approach until ILC adopted it for its rear-entry State Of the Art Suit of late January or early February 1965. ILC won the Apollo Block II Pressure Suit competition with a prototype also using such bearings made by Air-Lock. In 1968 NASA wanted upper-arm bearings that were more compact than Air-Lock could produce, so ILC took over the redesign and made the subsequent Apollo program upper-arm bearings in house.

17. *Shoulder Teflon Ferrule Multi-Directional Joint*—Invented by Hamilton Standard (U.S. Patent 3,492,672, Michael A. Marroni Jr., Douglas E. Getchell, and John J. Korabowski inventors), it was adopted by ILC for its 1965 Block II competition suit. Later, ILC switched from Teflon cord to plastic-coated aircraft cable in the joint.

18. *The Apollo Walking Brief*—ILC's Apollo brief system was invented by George Durney (U.S. Patent 3,699,589). This was initially developed under the Apollo contract and first appeared in November 1964. The brief was refined to become a highly successful system in ILC's Block II competition prototype in July 1965. With some subsequent refinements, this would be used for all the Apollo and Skylab pressure suits.

19. *Bubble Helmet*—In the 1940s, full-bubble helmets were popular with many manufacturers. Such helmets had acrylic bubbles and fell out of favor because switching from unpressurized breathing of ambient air to pressurized use necessitated donning and doffing the helmet. This led to development of flip-type pressure visors for high-altitude pressure suits. Acrylic also had poor impact properties. A broken Gemini (acrylic) pressure visor in 1964 caused

NASA to switch to polycarbonate pressure visors. Soon after this, NASA's Jim O'Kane and Bob Jones championed development of the polycarbonate full-bubble helmet. The problem with polycarbonate was the difficulty of making a full bubble while retaining optical quality. In 1967 the Air-Lock Corporation was the first organization to get optical quality helmet bubbles into production. Consequently, Air-Lock produced the O'Kane/Jones design for Apollo, Skylab, and Shuttle extravehicular spacesuits.

20. *Communications Carrier Assembly (a.k.a. "Snoopy Cap")*—During Apollo the speaker and earphones for the radio moved from being part of the helmet to being held on the head by a cloth cap. There were Hamilton, ILC, and David Clark versions. The David Clark Snoopy Cap was used on the Moon. For Shuttle and the International Space Station, NASA elected to use the David Clark design but have it manufactured by ILC.

21 *Liquid Cooling Garment and Multiple Water Connector*—Apollo manned testing in 1963 proved that gas cooling was impractical for extravehicular activity. This brought about two inventions. The first was the spacesuit's Liquid Cooling Garment (LCG) (U.S. Patent 3,289,748, David C. Jennings inventor). This garment provided both liquid cooling via water flowing through tubes that were in contact with the body and supplemental gas cooling via an open-mesh, tubing-constraint garment. The introduction of the LCG required cooling water to travel through the wall of the pressure suit. The second, in support of this, was Apollo's Multiple Water Connector (MWC), which was invented by Hamilton's Bradford Booker. The MWC allowed the LCG and the backpack life support system to independently connect and disconnect from the suit without breaking the pressure seal. The cooled water flowed from the backpack through the MWC to the LCG where the water flowed around the astronaut removing heat. The warmed water then flowed back through the MWC to the backpack life support system. The LCG and MWC were the result of Hamilton internally funded development. NASA purchased the patent rights to both in March 1966. NASA and ILC elected to allow Air-Lock to redesign the MWC in 1966.

22. *Helmet Drink Port*—While not shown on the helmet in Figures A2 and A3, a patent for "Helmet with Pressure Seal Port" (U.S. Patent 3,067,425, Russell S. Colley inventor) was filed on November 9, 1959. This helmet food and drink port was subsequently used by NASA on Gemini and Apollo helmets.

23. *Walking Ankle Joint*—Litton introduced the first pressure suit ankle joints in its Mk. I pressure suit of the late 1950s. This joint used hard gimbals to control pressurized movement. Hamilton Standard invented an all-soft boot with ankle joint for Apollo (U.S. Patent 3,605,293, Douglas E. Getchell, John J. Korabowski, and Michael A. Marroni Jr. inventors) which first appeared in the 1965 Apollo AX5H suit model. NASA's testing of the Hamilton-Goodrich Apollo Block II competition suit resulted in NASA desiring incorporation of a Hamilton-style ankle joint into the ILC Apollo suits that followed. As a result of schedule constraints, this was not accomplished until 1967.

24. *Rear Entry*—The rear-entry concept was introduced on a David Clark

pressure suit in the 1950s by designers Joe Ruseckas and John Flagg. This approach was refined by Dick Sears and Don Robbins for the prototype that gained David Clark the Gemini suit contract. The rear entry to the Gemini suits was so well liked by the Astronaut Corps that ILC started developing its own prototype under internal funding. It made its public debut in 1965 along with ILC's Block II competition suit. This was a feature on all Apollo 7 through 14 and 18 suits, as well as the Apollo 15 through 17 Command Module Pilot Suits.

25. *Pressure Sealing Zipper*—B.F. Goodrich developed and manufactured the pressure sealing zipper. Introduced in 1965, it made rear-entry zippers practical for Apollo. This significantly reduced gas leakage and in so doing permitted rear-entry suits to meet program requirements.

26. *Helmet Vent Duct*—The O'Kane/Jones ventilation duct was first used in their 1964 prototype full-bubble helmet and their subsequent prototypes. In 1967 the duct became part of the full-bubble helmets used in the Apollo and Skylab programs. NASA continued using this same helmet and duct system for Shuttle and International Space Station extravehicular spacesuit helmets right up to the present.

27. *Urine Collection and Transfer Assembly*—B.F. Goodrich were the first to develop such a system in the U.S. It was used on the Mercury program.

28. *Injection Patch*—The first such system developed in the U.S. was in the David Clark Apollo Block I program.

However, innovations are not the only way to view the origins of the Apollo spacesuit. The creation of the Apollo spacesuit was more than inventions. It was recognizing the value of the innovations, merging them into test proto-types, fully developing those designs, manufacturing the spacesuit elements with no room for error, making the deliveries to support the program time limita-tions and merging all the parts into a space exploration system that made history. This trail was complicated by the United States seeing the race to the Moon as a competition it could not afford to lose. When there was delay that could jeopardize having the lunar spacesuit in time to support the rest of the program, NASA funded a parallel, competing program to assure success. When one of the competing programs succeeded, NASA terminated the other. However, the parallelism and terminations resulted in a cross-pollination of knowledge and processes. This causes the Apollo spacesuit story to be very confusing to many. Thus, a map (Figure A4) leading to the successful develop-ment and use of the spacesuit is provided. Hopefully, it will be of help.

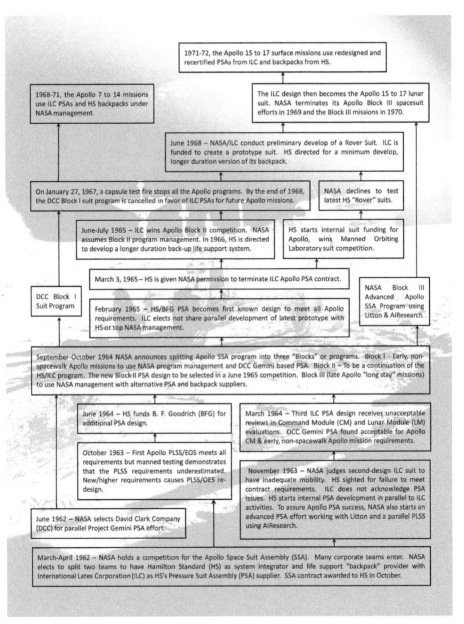

Figure A4. Map of key events.

APPENDIX II

Acronyms

CMP   Command Module Pilot (the suit type or model used by Command Module Pilots of the Apollo program)

CO_2 Carbon dioxide (a by-product of human respiration that when sufficiently concentrated becomes toxic)

EMU   Extravehicular Mobility Unit (the name given to the complete Apollo spacesuit system from 1965 to 1972)

EOS   Emergency Oxygen System (the name given to the Apollo Space Suit Assembly's backup life support system from 1962 to 1966)

EV   Extra-Vehicular (the suit type or model used by Lunar Module crewmembers)

EVA   Extra-Vehicular Activity (any activity in space conducted outside the spacecraft)

HS   Hamilton Standard Division of the United Aircraft Corporation (prior to 1975); Hamilton Standard Division of United Technologies (until 1999); Hamilton Sundstrand Division of United Technologies (until 2012); UTC Aerospace Systems (after 2012)

ILC   International Latex Corporation (from 1962 to 1969 ILC was headquartered in New York City with a spacesuit facility in Dover, Delaware; in 1969 spacesuit operations were spun off into a separate entity now named ILC-Dover LP; after Apollo, ILC-Dover formally moved to Frederica, Delaware)

ITMG   Integrated Thermal Meteoroid Garment (the outer protective cover garment attached to the pressure suit that was launched into and returned from space)

LCG   Liquid Cooling Garment (a device used to control the astronaut's thermal comfort by means of water flow through the PLSS)

LP   Limited Partnership as in ILC-Dover LP.

LiOH   Lithium hydroxide (used to remove Apollo carbon dioxide)

LSS   Life Support System

MOL   Manned Orbiting Laboratory

MWC   Multiple Water Connector (a system that allowed water to flow through the wall of the pressure suit when the LCG was being used and to provide a pressure seal when it was not being worn by the astronaut)

NACA   National Advisory Committee on Aeronautics (founded in 1915 to undertake, promote, and institutionalize aeronautical research in the United States; it was merged into NASA in 1958)

NASA   National Aeronautics and Space Administration (founded in 1958)

OPS   Oxygen Purge System (the name given to the Apollo spacesuit's backup life support system from 1966 to 1972)

PLSS   Portable Life Support System (the name given to the Apollo spacesuit's life support system backpack)

PGA   Pressure Garment Assembly (the name given to the pressure suit portion of the Apollo spacesuit system from 1962 to 1964)

PSA   Pressure Suit Assembly (the name given to the pressure suit portion of the Apollo spacesuit system from 1965 to 1972)

RAF   Royal Air Force (the U.K.'s air force)

SEVA   Stand-up Extra-Vehicular Activity

SSA   Space Suit Assembly (the name given to the complete Apollo spacesuit system from 1962 to 1964)

SI   The Smithsonian Institution (SI) is a group of museums and research centers administered by the Government of the United States, of which the National Air and Space Museum SI is a key part.

USAF   United States Air Force

UT   United Technologies (UTC Aerospace Systems was formed by the merger in 2012 of the Hamilton Sundstrand Division of United Technologies with the Goodrich Corporation newly acquired by UTC) .

UTC   United Technologies Corporation (before 1975 this was the United Aircraft Corporation, an American aircraft component manufacturer formed by the breakup of the United Aircraft & Transport Corporation in 1934)

APPENDIX III

Figure List

Note: These illustrations were provided by ILC-Dover LP. Permission to use the photos does not imply approval of this book or agreement with the contents of its story.